Ghost Road

Ghost Road

BEYOND THE DRIVERLESS CAR

ANTHONY M. TOWNSEND

W. W. NORTON & COMPANY

Independent Publishers Since 1923

For information about permission to reproduce selections from this book, write to Permissions, W. W. Norton & Company, Inc.,
500 Fifth Avenue, New York, NY 10110

For information about special discounts for bulk purchases,
please contact W. W. Norton Special Sales at
specialsales@wwnorton.com or 800-233-4830

Manufacturing by LSC Communications, Harrisonburg
Book design by Daniel Lagin
Production manager: Beth Steidle

ISBN 978-1-324-00152-2

W. W. Norton & Company, Inc., 500 Fifth Avenue, New York, N.Y. 10110
www.wwnorton.com

W. W. Norton & Company Ltd., 15 Carlisle Street, London W1D 3BS

1 2 3 4 5 6 7 8 9 0

IN MEMORY OF THE UNCLE I NEVER MET

William Patrick Herrschaft,
Petty Officer Second Class, US Navy,
killed in an automobile accident,
Patuxent, Maryland,
April 4, 1961

Contents

PART III

Taming the Autonomous Vehicle

Tables and Illustrations

Preface

There's one ancient story we all share, though the details differ from tribe to tribe. It speaks of a hero or god, who rides forth on a divine relic with a mind of its own. There are the flying carpets of *One Thousand and One Nights*, the collection of folktales from Islam's golden age. Even older Persian myths recount the feats of a great shah, Kay Kāvus, who flew to China on a magical throne held aloft by four eagles. But the airspace over the Middle Kingdom was already crowded. The most ancient stories of all tell of Chinese emperors who flew through the heavens on chariots and thrones.

Other vehicular lore is more down to earth. Slavs have terrorized generations of children with bedtime stories about the evil witch Baba Yaga, who roamed the Eurasian steppe in a monstrous hut propelled by two giant chicken legs. For the Welsh it was King Morgan the Generous, teleporting about Britain in a magic chariot. In the Nordic lands they worshipped Frey, the god of nature, whose self-steering ship could fold up small enough to fit in his pocket.

For thousands of years now we've looked on with envy at our deities' carefree journeys. Their enchanted artifacts embody our oldest ideas about transportation, our most ancient technological longings. We've never waited so long for a divine invention to arrive.

But behold! At long last, *we* will be the first human beings to achieve these immortal powers. Much as the internal combustion engine replaced animals at the yoke more than a century ago, supercomputers are now taking our place behind the wheel. Instead of spells and sorcery, a man-made magic we call *software* is finally making our ancient dreams real.

This computer takeover is transforming the familiar automobile into a new breed of vehicle—an *automated* vehicle, or AV. Our cars can already steer, stop, and park themselves with an astonishing degree of skill. But this *partial* automation is just the beginning of the delegation of driving duty to machines. With autopilots engaged, *fully* automated vehicles will soon begin rolling into our lives, first by the thousands, then by the millions, and then, around mid-century, by the billions. While we sleep, they'll haul the trash, patrol the streets, and do the other dull, dangerous, and dirty work of daily upkeep. When we wake, they'll deliver the goods we used to go out to shop for. And when we do sally forth, we'll step out like the wizards of yore. With muttered incantations, we'll summon our mounts. But instead of supernatural steeds, robotic runabouts will heed our calls.

And then . . . whoosh! Off we'll go. Just like the gods of old.

ON OCTOBER 9, 2010, artificial intelligence guru Sebastian Thrun posted a brief announcement to Google's corporate blog, revealing the search giant's clandestine effort to develop a fully self-driving car. No longer content to merely organize all the world's information, Google now aspired to orchestrate the world of transportation, too. Great things, Thrun told us, were right around the corner. "Our goal is to help prevent traffic accidents, free up people's time and reduce carbon emissions by fundamentally changing car use," he wrote. We were all too eager to believe.

Ten years passed. Private-sector funding for research on driverless systems surged tenfold, from $6 billion in 2015 to over $60 billion in 2018. AV prototypes logged millions of miles on real roads, and *billions* more cruising around inside virtual worlds—simulations nested inside simulations like so many Russian dolls. But a funny thing happened on the way to the future. The same companies that filled our heads with daydreams of self-driving cars now say that perfecting the technology will be trickier than they thought. Few of us have seen an AV, let alone ridden in one. Most of us doubt they'll ever truly be safe. And workers of the world tremble at the prospect of a job-destroying robovehicle invasion to come.

In the meantime, our attitudes about how we get around have changed. When Google fired the opening shot of this driverless revolution, using computers to fix the shortcomings of cars seemed like a sensible idea. But today, as dealers await shipments of the first self-driving SUVs, people everywhere are looking for alternatives to owning automobiles. We're riding mass transit and using for-hire vehicles more than ever. We've taken to electric bikes and scooters in droves. Automated vehicles are late to the party.

So perhaps we've hit this speed bump at just the right moment. The hype about high-tech cars has been laid on thick by self-driving soothsayers over the last decade. They've spun modern-day myths to replace the ancient ones, with predictions of a driverless future that's totally safe, traffic-free, and affordable to all. But the glitches in this meme are already starting to show. When GM's Cruise division tested AVs in San Francisco, for instance, its self-driving software sometimes failed to spot pedestrians and at other times braked suddenly after detecting phantom bicycles that weren't really there.

Our current view of what's to come is just as blinkered. That's why now is the time to raise the hood on this would-be driverless revolution. Because as those false promises are washed away, new futures will be revealed. Fantasies of highways filled with identical pod cars cruising in synchrony will fade. Visions of cities powered by computer-controlled vehicles of a thousand shapes, sizes, and speeds will take their place. We'll see that moving *people* isn't the killer app for AVs; moving *stuff* is. And horror stories about the clueless choices of computer chauffeurs will give way to the far greater danger of robot-powered transportation monopolies.

As we try to overcome our mistaken beliefs about the driverless future, we won't entirely escape our superstitions. That's because the way forward is haunted by specters both old and new. AVs will be our most intense and intimate encounter with artificial intelligence in the physical world, and we'll trust these mysterious new machine minds with our safety and welfare even after they've taken our jobs. More sinister spirits will watch us from the cloud, where bots that meter everything that moves will regulate our travels far more thoroughly than any government ever has. Most ghastly of all are the millions of restless souls—the future victims of avoidable crashes—who'll torment us if we don't embrace this technology with vigor.

I call this troubled future territory the *ghost road*. It doesn't show up on maps. But you'll find it soon enough, when you're cruising along in your computer car without a care. You'll look around and, for the first time, realize that you're now a stranger in a strange land, outnumbered and outclassed by software-steered machines.

THE GHOST ROAD is both beacon and warning. Despite delays and detours in the quest for vehicle automation, over the coming decade we'll finally master this technology. Computers and cars will combine, and neither will ever be the same again. The machines that result from this merger will transform how billions of people live, work, and travel in the century to come. But this is no repeat of the Motor Age's upheaval. AVs aren't dumb machines that follow a prescribed path. They will use artificial intelligence to make their way through the world on their own. And like animals they must be tamed, lest they overrun our human world and subjugate us instead.

If we succeed, the ghost road can be a friendly place for future generations. Some two billion motor vehicles could be scrapped, replaced by much safer and more energy-efficient substitutes. Hundreds of millions of people who struggle every day to overcome disabilities and isolation could become independent. We could rebuild our communities from the inside out by reclaiming the vast amount of land currently devoted to cars and commerce, and repurposing it for housing, parks, and plazas.

If we fail, the ghost road will instead become a prison, and AVs will hasten our decline. Eliminating the hassle of driving and slashing the cost of travel could send us onto the roads in unprecedented numbers, erasing our efforts to reduce traffic and fueling a new spike in carbon emissions. Governments would be bankrupted, as we replayed the wasteful infrastructure binge of twentieth-century sprawl on an altogether unimaginable scale. Tens of millions of professional drivers might lose their jobs, with no safety net or alternative livelihood to bail them out. And the poor would be left by the roadside, as the software agents of the ghost road lock out anyone with bad credit.

Taming the AV isn't simply a matter of getting the technology right. The real obstacles are our own institutions and intuitions—laws, business models, irrational fears, and old habits. And so, while this book closely examines the corporate strategies and public policies that will shape our prospects in the driverless revolution, it also looks at how we as individuals can make better choices in the days ahead. I'll share a set of principles, a personal code of conduct I call *big mobility*, that can guide your everyday decisions—whether you are a teenager or an industry titan. Think of big mobility as an ethos that will help you embrace the innovations to come while heeding the hard-knocks lessons about sustainable, inclusive mobility we've already learned.

You'll find this book asks as many questions as it answers. There's simply too much in play—entire industries, dozens of technologies, and huge and diverse populations—to predict how it might all turn out. But that is the most exciting part. The future is wide open for us to imagine something better. Now is the time to cast aside all we've been told and write the new myths of the driverless revolution to come.

PART I

On the Ghost Road

1   Fables of the Revolution

We tend to overestimate the effect of a technology in the short run and underestimate the effect in the long run.

—Roy Amara, Institute for the Future

We have so much to cover in an AV-powered future, there's little time to linger on the past. But we've been chasing this dream for a long time, and there's important history to take in before we dive into the latest advances. Let's start at the very beginning.

The first self-driving vehicles were ships. After centuries of wrestling with wind and waves, ancient sailors devised contraptions that harnessed these forces of nature to fill in for man. They were simple but ingenious solutions, like the sheet-to-tiller system, which is still used today (Figure 1-1). To rig it, you simply take the jib sheet (the rope that controls the smaller sail up front) and run it around a pulley and back across the deck. Finish by tying the bitter end to the tiller (the stick that steers the boat). Now, when a gust hits and the boat starts to round up into the wind, the jib will pull the rope around the pulley and yank the tiller, steering the vessel back the opposite way. Tricks like this helped clever

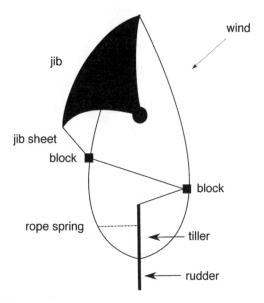

Figure 1-1. *Self-driving ship.* A traditional system for self-steering sailboats, the sheet-to-tiller rig.

mariners relieve the fatigue of long shifts at the helm during the Age of Sail. *You* can use it to crack open a cold one and enjoy the spray as your yacht plows through the whitecaps like a train on rails. And while tillers were repurposed to steer the first automobiles, this old technique didn't make the leap from sea to land—though we can imagine some frightful, fruitless attempts to make it do so. By 1891, the introduction of the steering wheel, by Benz, put the matter to rest.

On land, self-steering actually got harder when machines replaced animals. Motorization was a vast improvement over draft animals' muscle power, but the gain came at the expense of brain power. It had long been common for riders on horseback, and even cart drivers, to fall asleep at the reins. Their dutiful animals would simply keep following the road or stop dead in their tracks.

Cars and trucks, however, needed drivers to guide them second by second. Their soaring popularity, combined with the growing risks posed by their weight and speed, birthed a variety of experimental self-steering schemes. One 1925 demonstration of a remotely controlled vehicle in New York City offered a glimpse of driverless autos to come, simultaneously tantalizing and terrifying the public. Cruising down Broadway before thousands of onlookers, the optimistically named American Wonder drove "as if a phantom hand were at the wheel," reported the *New York Times*. In the 1920s, motor vehicles claimed tens of thousands of lives annually—a death rate 18 times higher than today. This new technology promised to render city streets safe once again. But those hopes were soon dashed when the futuristic vehicle's operators lost control—first at Sixty-Second Street and again moments later at Columbus Circle—before finally crashing the would-be wonder into another vehicle.

Despite this early misstep, the auto industry continued to daydream about remote-controlled cars. At the 1939 World's Fair, the *Futurama* exhibit by General Motors (GM) featured an enormous motorized diorama of an American city. Free-flowing highways plied by self-driving cars, trucks, and buses crisscrossed bustling districts of slender skyscrapers. There was even a "traffic control tower" where, the future city's designers imagined, dispatchers would direct the movements of tens of thousands of vehicles by radio. By the 1950s, guide wires embedded in the road surface had replaced radio as the preferred technology for remote-controlled vehicles. Ironically, it was RCA, the Radio Corporation of America, that staged the first successful demonstration of this approach in the 1950s.

These early prototypes showed the technical feasibility of automated driving, but their high cost and the lackluster demand for such features meant that neither radio-controlled nor wire-guided cars caught on. The price tag for guided-vehicle highways was thought to be as high as $200,000 per lane-mile. If fully built out, this road upgrade might have added more than 40 percent to the cost of building the Interstate Highway System, already the largest public works project in American history. Meanwhile, despite the dangers and drudgery of long or late-night drives,

automakers were still riding a wave of consumer *excitement* about driving. They focused on producing powerful new cars that were exhilarating to drive.

These early dreams imagined a self-driving future based on external guidance. But by the 1960s, the focus had shifted to harnessing the
new technology of computers to design vehicles that could truly, independently drive themselves *autonomously*, without outside help. At Stanford
University, for the first time anywhere, researchers built robots that used
cameras to see and computers to navigate. In highly controlled experiments, these early droids followed white lines and avoided obstacles
placed in their path.

Self-driving wasn't confined to the laboratory for long. CPUs and
image-processing techniques improved, so that by the late 1970s engineers at the University of Tsukuba's Mechanical Engineering Lab were
able to test the world's first self-driving passenger vehicle, on Japanese
roads. Traveling at speeds up to 20 miles per hour, these first AVs used
two video cameras to visually detect street markings. In the 1980s the
action moved to Europe, where Ernst Dickmanns, a professor at West
Germany's Armed Forces University, retrofitted a Mercedes-Benz van
with self-driving gadgets of his own design, launching a decade-long collaboration with auto giant Daimler. Finally, it was the Americans' turn, as
Carnegie Mellon University took the lead in the 1990s. As the competition
to build self-driving machines spread worldwide, the software improved
quickly and computers got ever faster, unlocking new possibilities. By the
decade's end, the first cross-country trips under automated control—in
the US, Germany, and Japan—were in the record books.

The most intense period of AV development was still to come. In the
early 2000s, the Pentagon took a growing interest in this emerging technology. To focus the efforts of scattered research groups and catalyze
stronger ties with the defense and auto industries, the Defense Advanced
Research Projects Agency (DARPA)—the US military's most independent
research-funding arm—organized a series of open competitions in 2004,

2005, and 2007. These "Grand Challenges," as they were called, offered millions of dollars in prize money and priceless prestige, and attracted dozens of teams from academia and industry. Putting their best hardware and software to the test, the competitors watched from afar as their AVs tried to traverse both open country and more suburban settings on an abandoned military base. The 2004 race ended without a winner—none of the entrants reached the finish line. But a year later, Stanford University's winning vehicle claimed the $2 million prize.

The DARPA contests accelerated the development of driverless vehicles. Stanford's first-place finish in 2005 was the result of its pioneering use of machine learning, an AI programming technique, in processing road imagery. But more important, the contests focused attention on the emerging technology's possibilities. No one was shocked by the military's rising interest in AVs. But it was the potential civilian applications that set off a sudden wave of speculation. For the first time, the practical commercial use of self-driving technology seemed within reach.

It was a wake-up call for the auto industry. But not everyone heard it. Most companies were preoccupied with the financial crisis of 2007–2008 and the global recession that followed. US automakers in particular were hamstrung when it came to capitalizing on the opportunity of AVs, which would require substantial further investment for the journey from lab to market. The automakers were going bankrupt or getting bailed out by the federal government. Instead, Silicon Valley moved forward. By 2009, the head of the winning Stanford University team, Sebastian Thrun, was leading a new self-driving-car project at Google. The search giant had bet big on Android, its highly successful operating system for mobile phones. Cars could become the next big computing platform, it seemed. Could Google stake a claim on the future of automotive software? It appeared to be a smart bet, bolstered by CEO and cofounder Larry Page's lifelong interest in AVs.

Google's move took a few years to sink in, but once it did, all hell broke loose—not only in the car business, but in the computer and cab industries as well. Suddenly, every major automaker, every ride-hail company, and competing cloudware giants like Apple hastily mobilized efforts to develop self-driving vehicles, too. When in-house projects failed to produce convincing results, many companies simply acquired promising startups to get hold of the needed technology instead. In a two-year period during 2016 and 2017 alone, some $80 billion surged into self-driving vehicle technologies. The biggest deal, Intel's panicked 2017 acquisition of computer-vision pioneer Mobileye, an Israel-based maker of computer-vision systems, was valued at an eye-watering $15 billion. As this flurry of mergers and acquisitions unfolded, the web of partnerships and cross holdings linking automakers and the tech sector grew ever more tangled. Two of the world's biggest consumer industries—computers and cars— had seen their future in each other. But they couldn't decide whether they wanted to get together or gobble each other up.

By 2018 the hard work and high finance had paid off. In December, Google spin-off Waymo quietly unwrapped the world's first truly self-driving taxi service, in Chandler, Arizona. More than 40 years after the first AV test-drive at Tsukuba, and nearly a decade after recruiting Thrun, the company started taking requests for driverless rides through the Phoenix suburbs. Reports said the tech giant had set aside more than $10 billion to build out its self-driving empire. At last, it seemed, the long and painful birthing of the AV was finally over.

The Promise of Automated Vehicles

"There is hardly a task that horse-drawn vehicles can do which cannot be done as well, and possibly better, with automobiles," reported the *New York Times* on January 12, 1903, as one of the world's first big auto shows opened its doors inside Madison Square Garden, then located at Twenty-Sixth Street and Madison Avenue. The *Times* was still at it a century later, this time hawking the engineering marvels of the self-driving age with a similar enthusiasm. "On my fourth day in a semi-driverless car," wrote columnist David Leonhardt in 2018, "I was ready to make a leap into the future." The paper of record isn't alone. Much like the automobile, AVs have unleashed bold speculation about the new technology's benefits to individuals and society. But what does that future promise?

First, self-driving technology can eliminate nearly all of the deaths caused by automobiles, say its champions. An estimated 60 million people were killed in motor vehicle crashes in the twentieth century. That's more than all of the military *and* civilian deaths during World War II. But even as cars have become much safer, the killing continues, as motor vehicles spread to new countries where skilled drivers and traffic regulations are in short supply. As auto use booms in China and India, more than 1.4 million road deaths occur worldwide every single year—stealing enough souls to fill a city the size of Dallas, Texas; Birmingham, England; or Kobe, Japan. The vast majority of these crashes would have been prevented with self-driving technology, advocates claim.

Second, AV boosters boast, traffic congestion as we know it will disappear. The economic toll of overcrowded roads is enormous, and is easier to measure than ever, thanks to location-tracking devices embedded in ubiquitous mobile phones. Using the vast troves of travel records these phones leave behind, telematics firm Inrix estimated that in the US alone, the cost of drivers' time wasted in traffic was over $305 billion a year, or nearly $1,500 per driver. The argument for AVs is that software-piloted cars can safely pack more cars closer together at highway speeds, thanks to faster braking reflexes. But AVs might also reduce some bottlenecks by simply spreading human populations farther apart, splaying settlements out over a wider expanse of land. When passengers in AVs can use travel time for work or leisure instead of keeping eyes on the road, the thinking goes, longer rides to less-congested areas won't be a bother.

Third, no one will be left behind by AVs, advocates hope. Cars expanded mobility for hundreds of millions of people in the twentieth century, but when the automobile's success dispersed the population and siphoned funds from mass transit, many found themselves facing new barriers to freely getting around. In the US alone, more than 25 million people have disabilities that limit travel—nearly one-sixth of the workforce. Not only will AVs bring automobile travel to those physically unable to drive, it is believed, they will open up new travel options for the very old, the very young, and those who can't afford cars of their own. As disabled people come off the sidelines and enter the workforce, as senior citizens get easier access to medical care, and as children enjoy access to a wider range of educational and enrichment opportunities, the social and economic benefits could be enormous.

When will this utopia arrive, you ask? Today AVs are still a novelty. Despite all the hassles, dangers, and drudgery of driving, *we* remain the most cost-effective "technology" suited to the task. By the time you read this, in the early 2020s, even if the wildest predictions come to pass, there will still be fewer than *one million* truly self-driving vehicles plying the world's highways, streets, and sidewalks. But AVs' numbers are destined to grow quickly as the decade rolls on. By 2030 the global head count of smart cars, trucks, and buses could creep into the *tens of millions.* They'll share the road with some two billion human-driven cars and trucks (give or take a few hundred million). Even then, it seems, AVs will be but a rounding error in the global population of automobiles. But the revolution will strike with surprise, surgical precision, and overwhelming force. As cyberpunk novelist William Gibson once famously said, "The future is already here—it's just not very evenly distributed."

The first changes we notice will occur in taxis. Most market analysts agree that *all* taxis in the industrialized nations will be automated by 2030. In the US, that's 300,000 vehicles. Add in all the Ubers and Lyfts and the total is closer to 1,000,000 in all. Swarming from our airports and resorts through our most beloved downtowns, driverless cabs could become the face of automation for a generation, and the gateway drug to driverless mobility for billions of passengers every year. The arrival of driverless cabs could radically change consumers' perception of cars. When computerized chauffeurs are a tap and a swipe away, and robotaxi rides are dirt cheap, people may opt out of auto ownership altogether. If we make the shift en masse, far fewer vehicles will be needed to move the same number of people that private cars do today.

But this silver lining may not come to be. Automation will also make private automobiles more useful, and software will radically reduce the hassles of ownership. Think about it for a moment. Automated cars will do more than drive for you—they'll also park themselves, take themselves to the garage for fuel and repairs, and pay their own insurance bills (with your money, of course). It's entirely likely that we'll simply swap our stupid cars for smart ones, and go on cruising around as we have.

In the long run we'll likely see a mix of both worlds. By 2040, even if shared AVs take over and new-car sales fall by 50 percent—a sea change, indeed—automakers will still be churning out some 30 million self-driving cars worldwide every year. Half will end up in China, another quarter in America, and the rest scattered across the EU, Japan, and emerging markets. Yet even as the business of making cars shrinks, the business of using cars—and vans, and scooters, and everything else that goes—will grow. What's left of today's $2 trillion global auto-manufacturing industry will be subsumed into a much larger market for "personal transportation services" that's projected to reach $7 to $10 trillion a year by mid-century, roughly the size of the entire EU economy today. Waymo alone wants to capture a $1.7 trillion annual share by 2030. But Uber, Amazon, and Alibaba—not to mention Ford, GM, and VW, among others—aren't ceding this new frontier without a fight. They have their own designs on the service businesses of the self-driving future, too.

So while the driverless revolution starts with a trickle, before long that slow drip will become a torrent. By 2050 or thereabouts, most human-driven cars will be gone. A smaller, smarter fleet of self-driving vehicles of many shapes and sizes will have replaced them. Some will be private, some will be shared. Some will move a single person, some will haul a hundred or more. Many won't carry anyone at all, and instead will busy themselves with shuttling around an unceasing flood of goods unleashed by the triumph of online shopping. Some will help us by simply watching over our urban world or directing traffic. All told, our diverse fleet of AVs will log vastly more miles than our cars do today.

It's tempting to see the driverless revolution as a repeat of our twentieth-century experience with cars, only on a larger, computer-choreographed scale. But nothing in our past can prepare us for what lies ahead. At full tilt, the pace of change will bewilder us. In the US, full *motorization* took about 60 years—from roughly 1920, when cars started arriving in cities in large numbers, to 1980, when metro areas everywhere started to choke on their vast numbers. The next 40 years, from 1980 to 2020, was a period of *saturation*. The average number of hours spent in traffic by commuters nearly tripled, and the economic cost of traffic congestion grew tenfold, to $166 billion annually. We have spent much of this time seeking ways to curb auto use and invest in alternatives. But *automation* could play out in as little as 20 to 30 years—the span of a single generation.

Self-Driving Suburbs and Car-Lite Communes

If our history with the automobile does teach us anything—it is that the future we find in the driverless revolution won't be the one we expected.

Consider, for instance, that Henry Ford originally built the Model T for farmers. The car was cheap, rugged, and simple to repair. Indeed, it was a huge success in rural areas and connected farmers to urban markets. But it was city dwellers and a new suburban middle class who soon turned the new machine to their own purposes. Ford's neo-Jeffersonian vision of a nation of mechanized farmers gave way to a metropolitan reality—private cars carried millions more commuters to factories and offices instead.

Similarly, trucking helped spread factories deep into the countryside, a development that caught small towns by surprise. This sudden wave of industrial sprawl, which preceded the residential kind, was the impetus for the adoption of municipal zoning laws across America in the 1920s, which put strict controls on land use for the first time. In the process, thousands of communities adopted low-density layouts that separated housing from jobs, inadvertently baking in dependence on automobiles for generations to come.

Experiences like these highlight why predicting the impact of transportation innovations is risky business. When we try to map out how millions of people will put new transportation technologies to use for business, household chores, and leisure; how those small-scale behavioral changes will add up over time to bigger systemic changes; and how public institutions and political factions will react—we're doing little more than guessing. But while we cannot *predict* the future, that doesn't mean we shouldn't try to *forecast* it. On the contrary—by anticipating what plausibly *could* happen, we develop skills that will reduce our shock and help us recognize the future we eventually do face. By thinking through the possibilities, we can also take actions that prepare us for many different scenarios, not just one.

Unfortunately, much of today's speculation about the driverless revolution does just the opposite. It takes the vast potential to rework our world with automated vehicles and boils the futures-to-come down to just two possibilities.

Let's call the first version of the driverless future *self-driving suburbs.* This is a vision where we all wake up with an autonomous, electric vehicle parked in the garage of our solar-powered, suburban home. It offers us convenient, consumer-friendly mobility choreographed by computer. This is the future peddled in one shape or another by Tesla, Google, and GM.

Self-driving suburbs are the twentieth century, turbocharged. They promise a world where individuals have almost unlimited mobility at their fingertips, with few of the costs or risks associated with private cars today. Traffic congestion is rare, and hardly a bother when it happens. Crashes are unheard of, except for the occasional, colossal late-night robot-on-robot interstate pileup. Travel becomes more like following a hyperlink than embarking on a journey. You click where you want to go and then space out for a while until you pop up somewhere else, swiping or sleeping away the hours in between. "Drive until you qualify," a home buyer's mantra for the trade-off between cheaper housing and a longer commute, takes on a whole new meaning when software's at the wheel and the exurban frontier marches ever farther into the hinterlands.

The other school of thought shifts our focus from the periphery to the center. Let's call this vision *car-lite communes*. This is the preferred future of many mayors, architects, and activists. A growing number of companies are buying in, too, mostly those who think the big money in the self-driving revolution will be made selling services like ride-for-hire and delivery instead of vehicles. This strategy is all about exploiting AVs to pack people more closely together instead of spreading them apart—and in so doing, turning cities into green machines that are safe, affordable, and healthy for all.

Car-lite communes offer a collectivist counterpoint to the capitalist manifesto of self-driving suburbs. Following this vision means throwing out much of the transportation system of the twentieth century and replacing it with one redesigned along the lines of the cloud. Private cars are banned, replaced by fleets of shared electric, autonomous taxis, one-tenth in number but just as convenient and much less expensive to ride in. Every inch of every street is electronically tolled so that we all pay a fair share, instead of allowing road hogs to dominate this common space for free, as they do today. Walking and biking soar in popularity once the streets are cleared of roaring traffic, and there's lots of room left over for new parks and plazas. What vehicles remain are all automated, and either avoid busy crossings or timidly pick their way through the crowd.

These scenarios are designed to sell us, like pilot episodes of future worlds to come. At their best they inspire us to seek change on a massive scale. At their worst they become deeply misleading, propaganda for a driverless revolution their proponents would like to bring about. They present the future as a choice between right and wrong, where ideologies that explain the world as it is today are dressed up in the costume of AVs.

These visions also leave out what doesn't concern them, and pave over inconvenient likelihoods. Proponents of self-driving suburbs show little concern for the proven damage of sprawl—social isolation, childhood obesity, and wasteful use of energy. They forget that many urban areas face geographic barriers to expanding farther outward, and overlook the value of preserving remaining open space in its natural state. But car-lite communards have a big blind spot, too, when it comes to sharing. Even today, in the hipster enclaves of Brooklyn and Oakland, shared taxis account for only 40 percent of trips. But to achieve a truly transformational clearing of roads in the AV age, studies show 80 percent is needed. One think tank suggests "a range of strong policies to achieve" the needed uptake. But how much arm-twisting are we talking about? Will shared taxis be foisted upon the poor, only to clear the roads for the cars and cargo of the rich?

On top of their own contradictions, both narratives leave out too many of the twists and turns that will actually shape the future. Where are the failures and false starts that will litter the trail of any truly disruptive technology? And while the winners are clear enough, who gets screwed by the spread of self-driving? Crucially, both gloss over the all-but-intractable issue of data. While future AVs will mostly be electric, in lieu of tailpipe gases they'll leave behind a new form of toxic sludge—a sensor trail that tells all about us, where we've been, and who we've met along the way.

It's clear that neither of these visions is ready for prime time. But despite the limitations of both, throughout the book I'll use car-lite communes and self-driving suburbs as foils to test your assumptions and gauge my own. Meanwhile, we will roam further afield in constructing a new outlook on what's to come. Along the way, we'll explore how policy, markets, and human factors—not just engineering and design—will ultimately shape the driverless revolution's big changes: whether it's cheaper to own a car or share one; whether it's better to live close together or far apart; and whether you pay a company, a government, or a fellow resident to cart you around.

The Road Ahead: Three Big Stories

Three big stories—the new fables of the driverless revolution—will steer our journey along the ghost road. These three stories explain how self-driving technology will set changes in motion that reshape our world. And they zero in on the choices we have to make—as consumers and citizens, workers and entrepreneurs, and policymakers alike—which will mean the difference between one outcome and another.

The first big story is *specialization.* For nearly a century, engineers and entertainers have painted pictures of a self-driving future. It's always the same—a world of highways filled with identical pod cars, gliding along in perfect synchronization. This vision is so prevalent, it seems inevitable and is rarely questioned. It still shapes consumer expectations, corporate strategy, and technical standards for the self-driving future. Yet it's a world largely dreamed up by the very control freaks who're inventing AV technology.

Instead, in the driverless future, vehicular variety will flourish. Passenger cars dominate the roads today—but bikes, buses, trucks, and taxis are growing in number. Already, they are where the most exciting innovations in transportation technology are taking place. And automation will make all of these types of vehicles better. Their advantages over cars will grow and their disadvantages will diminish. Meanwhile, entirely new kinds of AVs are being dreamed up every day. We'll build vehicles that are bigger, smaller, slower, or faster than ever before—from self-driving shoes to autonomous buildings. These new craft will do things cars never could. We'll find them strange, yet surprisingly useful, and delightful too.

The second big story, *materialization*, builds on the first by shifting our focus from passengers to freight. Right now, we spend too much time thinking about how the way *we* travel will change in the self-driving age—but moving people will soon take a back seat to moving stuff. E-commerce is already driving a historic surge in shipments into local communities, businesses, and homes. Automation will lubricate this retail revolution. And tectonic shifts will follow—in how we shop, where and why we travel to do it, flows of waste and recycled materials, and the prospects for small businesses and job seekers, too.

The third big story is the *financialization* of mobility. You'll find little in this book about the safety of AVs. It's a red herring. We'll either perfect self-driving technology, or there won't be an AV industry to speak of. Instead, a much greater regulatory crisis looms down the ghost road. As automation hooks up local transportation and global capital in lucrative new arrays, our concern will shift to the wild dynamics of vast new mobility markets.

This is dark and uncharted territory. When every movement is precisely tracked, future demand for taxis and trains will become the target of speculation. Imagine new financial instruments, securitized by the highly predictable revenue streams produced by AVs. Yet when AVs and trading algorithms interact on a massive scale, unforeseeable and violent market swings may flare up. And who will wield this new financial power, and who does it put at risk? Much as Uber's introduction of surge pricing showed, the expansion of financial innovation in the mobility sector will deliver highly unequal benefits. What's worse, as the power of mobility financiers grows, the very same tools that cities will deploy to manage the AV invasion—such as congestion pricing—may be co-opted by speculators, weaponized, and turned back against local governments.

Each of these three big stories breaks down a popular yet mistaken assumption about the future of autonomous vehicles, laying out vital misconceptions and gaps in our knowledge. From there, we build up a new understanding of what's really going on and how it could play out in the coming years. Taken together, the three stories point to a future that's entirely different from anything we've yet imagined. It is a world where AV innovation is riskier but there's far more of it and the potential payoffs are much greater. It is a place where the transportation infrastructure we have no longer meets our needs but it isn't yet clear what must replace it or how to pay for it. And it foretells an age when the opportunity is greater for big economic leaps but they may come at the cost of still wider inequality.

The three big stories—specialization, materialization, and financialization—focus our attention on currents of change that will push everything along over the coming decades. If you try to map the likely path of every technology, track every fast-growing company, and predict every permutation of the future you'll go mad. These details are all but impossible to anticipate, their interplay infinitely more so. The trends underpinning these three stories, however, will outlast any one company, invention, or event. As we explore each story in turn, don't think of them as bold predictions. Rather, consider each one a working hypothesis about which way the driverless revolution may break and what's at stake when it does. Embed them in your thinking and all your forecasts will be future-proof.

The rest of this book explores the three stories, revealing connections that tie them together into a comprehensive outlook on the driverless revolution. Part I, "On the Ghost Road," covers the most confident territory. Here, AVs are mostly still serving us. We'll see how AVs' remarkable specialization can bring about an explosion of useful variety—not just in vehicle designs, but in the services they perform, and in the nature of driving itself. Part II, "No Man's Land," moves out of our comfort zone, into spaces where machines have more sway, to explore the logic of automated delivery. This materialization of online commerce is potentially the most destabilizing economic development of our lifetimes, with broad-reaching impacts for retail, local services, transportation, and land use. We'll explore the difficult choices ahead that will determine whether it turns us all into inveterate couch potatoes or carbon-negative planet-savers. Finally, in Part III, "Taming the AV," the gloves come off as human and machine destinies come into conflict. We'll examine the financialization of fully automated mobility and its physical impacts on communities—and get drilled on a new personal code for living well in the driverless age.

As we go along, you'll see how much is up for grabs in the driverless revolution. Even if all it upends is the auto industry, the arrival of AVs will throw into disarray a global business that employs millions, touches the lives of billions, and generates trillions of dollars of commerce. But the changes already in motion will reach far beyond the manufacture of cars. These changes will thoroughly remake taxis and public transit. And they will transform the places people use AVs to travel to—stores, health-care providers, and schools. There is no area of social and economic life that will escape the AV's disruptions.

The driverless revolution also entails huge risks, which I'll take pains to lay out. AVs could unleash more travel—worsening traffic problems instead of solving them. Automated freight might decimate local commerce, sapping communities of activity and badly needed entry-level jobs. The injustices and destabilizing impacts of financial markets that have rocked housing, energy, water, and food may be felt in transportation too.

And yet the ghost road may pose an even more fundamental challenge. AVs will be our first long-term, large-scale test case of cohabitation with artificial intelligence, the fast-advancing crop of software that seeks to replicate, or at least replace, human cognition. As thinking machines leave the confines of the factory, the test track, the lab, and the battlefield, will these robotic minds and their makers learn to play by our rules—or demand that we adapt to their limits instead?

I'm apprehensively optimistic. And you should be too. If used wisely in the years ahead, AVs could be one of the most powerful tools we have for reducing carbon emissions and halting global warming. Just as the automobile did, AVs provide an opportunity to rethink our world. We have the chance to shape this technology to fit into new designs for permanently viable communities where everyone has access to good homes, good jobs, and green space. And if that doesn't work, we'll enjoy a few decades of the most fun human beings have ever had on wheels.

For better or worse, one thing is now clear. The self-driving genie has escaped from its bottle. And it will not easily go back in.

2　Deconstructing Driving

In future commutes, you won't have to focus much on driving anymore. You'll be able to have hot pot and sing karaoke on the ride.

—Robin Li, CEO, Baidu

In many parts of the world, the automobile is what turns children into
adults. As their 16th or 17th birthday approaches, eager teens borrow some wheels and prepare for the road test—the last barrier that
stands between them and boundless freedom. But mastering the skill
of driving at such an unripe age is hard. One must overcome the erratic
impulses of a developing brain and body to precisely guide a two-ton
machine that harnesses the power of a hundred or more horses. The
examiner may choose any road and demand performance of a variety
of complex maneuvers (parallel parking!). Rain or shine, day or night—
aspiring drivers must be prepared to deal with any conditions nature
may throw at them.

For those who make the grade, the privilege of driving provides a
thrilling sense of new possibilities. But it also brings the burden of vast

new responsibilities. For as we memorize the rules of the road, we also take responsibility for the severe consequences of breaking them. We learn to make split-second life-and-death decisions that affect not only ourselves but the well-being of our passengers, pedestrians in our path, and other motorists sharing the road.

Cars teach us to make tough financial choices, too. Some hit close to home. Borrow the family car or save for one's own? Get a part-time job to pay for insurance and gas? Other questions seem almost too far-off to feel. How does the price of gas on the corner reflect tensions in the Middle East? What is the carbon footprint of my commute? Our daily decisions about automobiles make amateur economists, political scientists, and environmentalists of us all.

Today, this coming-of-age story is changing fast. Cars no longer play the leading part they once did. In the 1980s, about 70 percent of 17-year-olds in America had their driver's license. But by 2014, fewer than 45 percent did. Kids didn't ditch driving only in the US, however. A similar slide shows up in Canada, Great Britain, Germany, Japan, South Korea, Sweden, and Norway.

What happened? One plausible explanation is the most obvious—the sudden and sweeping arrival of digital technology on the teen scene. By 2018, more than 90 percent of American teens owned or had access to a smartphone, and almost all had a computer or game console at home. It didn't take a rocket scientist to see that teenagers had their faces buried in phones and tablets far too much to take driving seriously for long. What's more, a new breed of helicopter parents was all too eager to enable them. As *The Atlantic* reported in 2017: "For some, Mom and Dad are such good chauffeurs that there's no urgent need to drive. . . . Teens today described getting their license as something to be nagged into by their parents—a notion that would have been unthinkable to previous generations."

When I first heard this theory a few years ago, I immediately related. Because it was my story, too. As a teenager in the early 1990s, I was happier on my skateboard than inside a sedan. When my 17th birthday arrived,

I spent my money on a computer rather than a car. Back then I was an oddball, but today I would fit right in.

Many observers jumped to a similar conclusion—that lots of teens were swapping cars for phones. But the timeline didn't make sense. When smartphones hit the streets in big numbers in the mid-2000s, the teen turn away from driving was already in full swing. The decline began in the 1980s, after a widespread shift to *graduated licensing*, a more stringent scheme for permitting new drivers. These reforms sought to ease teens into life behind the wheel, phasing in driving privileges over time. Under the new rules, prospective drivers were typically required to keep detailed diaries of supervised time behind the wheel. Not surprisingly, many teens simply waited until age 20, when the restrictions no longer applied. So, smartphones didn't *cause* the decline in teen driving. In fact, the rate of decline in licensing was already slowing down when the iPhone went on sale in 2007.

Smartphones are, however, playing a central role in how teens cope with the more restrictive world they're growing up in. Gadgets in hand, young people are using their electronic powers to reorganize the entire automotive system around their own needs and desires, on their own terms. And who can blame the taste of a new generation? When you discover that hiring a taxi is as easy as ordering a pizza or finding a date, why bother with the hassles of cars? For teens, a car is something to be summoned with a few swipes. The endless search for parking, monthly loan payments, and smelly gas pumps? Leave that to someone else.

Driving isn't dying because of technology. It's simply falling out of fashion. In dealer showrooms across the land, the result is undeniable— the typical American car buyer today is more than 50 years old.

THIS MAY BE the first generation to carry car-summoning supercomputers in their pockets. But it isn't the first time we've turned to technology to reduce the cognitive demands of driving.

By the 1950s, Americans began to discover that the chore of driving could prevent them from enjoying the good life. Instead of being liberated by the automobile, our scarce spare time was spent shackled to the steering wheel. Driving, it turned out, got in the way of popular pastimes like sleeping and drinking.

Tales of late-night encounters between policemen and incapacitated drivers quickly became part of pop culture. One of the first reality shows, CBS's *Could This Be You?*, was broadcast out of Seattle in the 1940s to a massive Sunday evening radio audience all across the Pacific Northwest. Listeners would hang on every word of the recorded roadside interrogations, a half century before *Cops* tantalized television viewers. "At times the officer's voice could be heard above the wail of the siren, surmising that the driver was probably intoxicated," recounts one chronicle, but "even then, sometimes the driver was just weary from a day's work or too many hours behind the wheel."

Some relief—for the weary, if not for the sloshed—came a few years later, thanks to the efforts of Ralph Teetor, the president of Indiana piston-ring maker Perfect Circle. During World War II, the US had imposed a 35-mile-per-hour national speed limit to curtail civilian gasoline consumption. Inspired to aid in the war effort, Teetor invented an automatic throttle control he called the Speedostat. After patenting the device in 1953, Teetor licensed it to carmakers, who coined their own ungainly monikers, including Controlmatic, Touchomatic, and my personal favorite, Pressomatic. But within a few years, GM's Cadillac division devised a more approachable name that would come to describe them all— Cruise Control.

Cruise control promised many of the same benefits as today's AVs. As advertised, the gadget provided some relief from the physical and mental fatigue of long periods of foot-pedal throttle control. And by maintaining more constant engine speed, fuel savings of 15 percent or more could be achieved. But, foreshadowing growing concerns about partially automated vehicles coming to market today, cruise control has made highway driving *more* dangerous. A 2013 French study involving close observation of 90 men and women in a driving simulator found that sloppy, risky, and drowsy driving occurred *more* frequently when cruise control was switched on. A more recent MIT study, which analyzed telemetry from more than 270 million car trips, found that drivers spent as much as 10 percent of the time that cruise control was activated sending texts and emails or using apps.

Long before cruise control's liabilities came to light, more fundamental safety lapses by automakers triggered a wave of regulation and government oversight. The effect on automation efforts was chilling. Gone were the gimmicky names and expansive visions of self-driving sedans. As consumer-protection laws expanded in the 1970s and 1980s, carmakers focused instead on behind-the-scenes automation that didn't directly involve drivers. These new features sported reassuring names packed with engineering jargon—an alphabet word soup meant more to intimidate than to inform. There was ABS (antilock braking), ESC (electronic stability control, to reduce the risk of rollovers), and TCS (traction control system, an antiskid feature).

This timid terminology, however, has outlived its usefulness. We've reached a rhetorical crossroad as car and computer combine. The old labels no longer apply, and new ones must take their place. The hunt is on for a new lingo to describe the high tech of tomorrow. But in a world crowded with new technology, carmakers' pitch for automotive automation is shifting back to convenience and comfort, just as in the days before crash-test dummies became public celebrities. And it's these new words, perhaps more than anything else, that will shape our hopes, dreams, and fears on the journey ahead.

Caching Attention

Of all the terms used to describe the capabilities of the coming generation of AVs, none gets as much mileage as *self-driving*. And no one has wielded it with more moxie than Tesla, the Silicon Valley automaker. Elon Musk's wonder cars redefined the electric vehicle by making it sporty and sexy. Now, the company is doing the same with automated driving. But a more conservative spirit prevailed when it came time to name the new feature. Tesla's Autopilot feature follows in the footsteps of Chrysler's Auto-pilot, a Teetor-designed cruise control introduced in the 1958 Imperial.

I got my first look at Tesla's Autopilot a few months after its 2014 release, in Oslo, of all places. The Norwegian government has long encouraged the purchase of electric vehicles (EVs) with generous tax credits, and local authorities allow EVs to drive in car-pool lanes. Norway, one of the world's richest nations, is eager to ditch fossil fuels (despite its long exploitation of North Sea oil and gas). As a result, the capital city's rush hour is an odd sight—traffic jams of rich, old, white guys driving alone in their pricey Teslas.

Sliding in behind the wheel of a sleek black Tesla and going with the flow of Autopilot feels a bit like driving an iPad. In the center of the dashboard is a computer-generated version of your car, shot (as in a driving game) from a chase helicopter's point of view. Two icons, a speedometer and a steering wheel, turn from gray to blue to indicate that the computer has taken control of these functions. Pie-shaped polygons project from the corners of your vehicle, indicating the detection of threats or obstacles. Nearby vehicles pop up on the screen as you pass or are overtaken by them, the computer doing its best to approximate their size and shape. Autopilot's audience for this little digitized drama is obvious—the phone-toting teens of today, who'll be the car buyers of tomorrow. But there is something for the geezers, too. Even as this confident car looks forward to a fully computer-controlled future, it also wants to recapture the initial excitement around cruise control and revive the Space Age's untainted faith in technological progress.

By 2016, the marketing for Autopilot had kicked into high gear. "Full *Self-Driving* Hardware on All Cars," the company promised in a new campaign. But Tesla played fast and loose with the meaning of this loaded term. It hadn't yet created the software to actually do it. In its then-current state, Autopilot was little more than a well-coordinated mash-up of existing automation features—adaptive cruise control, automatic emergency braking, and lane-change assist. Unlike AVs developed by other automakers, Teslas rely on a significantly less elaborate sensor system. In lieu of the typical rooftop laser rangefinder used by most AVs, Autopilot "sees" through a much less costly array of digital cameras (eight in all) and a dozen radar and ultrasonic sensors. While other AVs are guided by detailed three-dimensional scans of the road ahead, Autopilot deduces the terrain the way humans do, by synthesizing separate two-dimensional images with slightly different viewpoints to create a sense of depth.

It's frighteningly easy to discover Autopilot's limits. In 2019, I took another Tesla for a spin through Park City, Utah, this time as a passenger. An old friend, Terry Schmidt, showed me the features of his new Model S as we cruised along Utah State Road 224, approaching the entrance to the Utah Olympic Park, built for the 2002 Winter Games. We followed a FedEx delivery van, as Terry toggled Autopilot's setting for following distance down to 1, the most aggressive level. Even at 55 miles per hour, the car inched up tight behind the vehicle ahead. By the time I noticed how closely we were tailgating him, the van had signaled a turn and veered into the rightmost lane. Sensing the open road ahead, the Tesla poured on the torque, instantly bringing us back up to speed. A traffic signal, now 150 feet ahead, turned from yellow to red.

"They haven't released the red-light-detection feature yet," Terry explained, slamming the brakes and bringing the car to a quick halt. Crossing traffic zipped across the intersection—right where we *would have been*, had Autopilot had its way. This sequence of events isn't unusual. Autopilot routinely lets drivers get themselves into equally unpredictable and dangerous situations. Tesla owners have also reported a startling number of bugs. Most worrisome are the ghost cars—false positives that live only on the screen—as well as the even scarier instances when there's really a car there next to you but the screen shows a void. Three fatal Tesla crashes between 2016 and 2018 all involved collisions with objects that were apparently mistakenly classified by Autopilot.

As we waited at the light, I caught my breath. Unfazed, Terry continued the tour of the Tesla's features, flipping through the car's on-screen menus. He selected one of the newest features, "Navigate on Autopilot," which automates lane-changing decisions on highways to maintain a target speed. It has a control that allows one to fine-tune the computerized driver's aggression level in three steps. There's "mild," for those still getting used to self-driving software; "average," which provides a suitable default for the masses; and the most reckless notch . . . "Mad Max," named after the 1979 Mel Gibson film about a postapocalyptic wasteland ruled by renegade road gangs. This snarky moniker seems clever at first. But it's a tone-deaf move for a company whose software already has a body count.

Autopilot's implication in fatal accidents has focused scrutiny on how the software perceives the outside world. But it is the system's shortcomings in tracking what's happening inside the vehicle that poses the biggest short-term challenge to self-driving technology's reputation. Take the case of Joshua Brown, who in 2016 became the first person to die in an Autopilot-involved crash. According to federal investigators, Brown spent a mere 25 seconds of the final 37 minutes of the trip with his hands on the steering wheel. Before this crash, Autopilot was designed to wait as long as five minutes under high-speed, straight-road highway driving before issuing an initial visual warning to the driver. A single light touch was all that was needed to dismiss the alert. Under this scheme, it was easy for drivers to become deeply engaged in other activities for long periods of time even while periodically satisfying the wheel-touch timer's demands.

Design changes made in the wake of Brown's death significantly tightened Autopilot's warning window. Autopilot now sounds an audible alert after 60 seconds of hands-free operation, and disables itself entirely after three warnings in the same hour, resetting only after the car is parked.

But are these new restrictions enough? Autopilot continues to produce a disturbing chain of incidents of driver malfeasance. In 2018, a Tesla owner was reported riding in the passenger seat while Autopilot steered along a UK highway. A few weeks later, the California Highway Patrol was forced to execute a two-car pincer maneuver to wrestle an Autopilot-controlled Tesla to a stop on US 101 outside Redwood City. The intoxicated driver was found passed out behind the wheel. Watchdog group Consumer Reports, which conducted an exhaustive survey of partial-automation systems in 2018, found Tesla's approach to keeping the driver engaged deeply lacking, giving it the lowest possible grade. "Because of the impressive ability of Tesla's Autopilot to keep the vehicle centered in its lane," the reviewers complained, "it's easy for drivers to become over-reliant on it."

The warning lights about partially automated driving have been blinking for years. In 2010, researchers at Virginia Tech placed subjects in simulated partially automated AVs for a three-hour road trip. When the scientists switched on the test vehicle's lane-keeping software—a far more limited form of self-driving than Tesla's Autopilot—fully 58 percent of drivers watched a DVD and another 25 percent did some reading. Some drivers spent as much as one-third of the three-hour trip looking away from the road.

Results like these are why old-guard automakers are taking a far more cautious approach to attention. GM's Super Cruise, introduced in 2017, sports a full-fledged driver-surveillance system, with an infrared head-tracking camera "pointed at the driver's face." The system starts nagging if you look away for more than five seconds. Keep ignoring it, and after 15 seconds it disengages. If Tesla is an absentminded babysitter, Super Cruise is a helicopter parent.

I don't envy the designers of Autopilot and Super Cruise. Making partial-self-driving technology both roadworthy and appealing to car buyers isn't easy. In aviation, an entire science of "crew resource management" emerged in the 1980s to reduce the risks of heavy cockpit automation. Comparatively speaking, automakers are just getting started. Yet until full vehicle automation can be achieved—and computers relieve us entirely of all driving tasks—our eyes, ears, and minds must be managed as carefully as torque, throttle, and traction.

―――

IN THE WAR for consumer attention, cars have been losing ground to screens for a long time. In the US, the amount of time people spend traveling peaked in the 1990s. Today, the average American spends 22 percent less time running around on a weekly basis than a generation ago. And how we spend the time we aren't on the run has shifted—away from work and chores toward home-based leisure activities like television, computer use, and sleeping. If you have ever observed teenagers for any length of time, you won't be surprised to find that their demographic drives much of this trend.

That could all change quickly, however, as couch and computer merge inside the belly of fully self-driving vehicles. Automakers are already reimagining vehicle interiors as living rooms and offices on wheels. Audi's concept car of the future, the Long Distance Lounge, features a massive window that doubles as a screen—mirroring content from occupants' mobile devices or overlaying information on views of the outside world.

Digital amenities like these—and the peace of mind to enjoy them while computer chauffeurs watch the road—will be a welcome change. People waste a lot of valuable time in cars. It's estimated that by 2050, globally, self-driving vehicles could free up 250 million hours that are wasted commuting each year in the world's most congested cities. That would be worth $150 billion in the US alone, where 86 percent of the workforce commutes by private automobile. Putting back to work all those idle brains stuck in traffic promises to be one of the biggest benefits of full vehicle automation.

There's just one catch. There's little evidence that people will actually use their newfound time to work. Recent surveys in America, Europe, and the UK reveal the opposite—people have a strong aversion to in-car labor. When potential AV buyers are asked how they expect to spend their saved time, the top responses include watching movies, chatting with friends and family, surfing the internet, looking out the window, and sleeping. Work is at or near the bottom of the list. These preferences appear durable, too. They've changed little across a number of surveys over the last five years, despite much wider awareness of self-driving technology's actual capabilities.

This reality may take some time for auto designers to adjust to. Work is so often assumed to be the main focus of future life inside AVs, it has become a cliché. As one tech blogger put it, panning a collection of AV concept cars, "Why do all of these interior designs look like miniature conference rooms? No matter how well a car drives itself or how confident you are in looking away from the road, the last thing this world needs is more conference rooms."

On the other hand, the digital-media business is already gearing up to shape this new realm of the mobile digital experience. Audi has recruited Disney to design in-car virtual-reality experiences for media-savvy passengers. Kia built a concept car that uses facial scans to identify and read its occupants' emotions to fine-tune content recommendations. And the market for in-car media may be effectively limitless. Collectively, passengers in AVs will be the biggest, wealthiest captive audience ever assembled. With a 360-degree wraparound wall of screens, fat pipes for content piggybacked on car computers' uplink to the cloud, and nothing for passengers to do for hours on end, in-car entertainment may one day be a bigger market than the entire auto industry today.

It's here that automakers' investments in in-car surveillance will pay off most. Although driver monitoring seems like a stopgap while partially automated vehicles rule the roads, once computers permanently take the wheel, those cameras aren't going away. They'll be repurposed to scan you and serve up precision-targeted media. There are even sensors on the way that uniquely identify your heartbeat—under the guise of measuring driving stress.

One likely business model for the self-driving age, then, is the capture and monetization of our attention. GM already tracks what a driver listens to on the radio and shapes each ad pitch by combining that data with records about where the driver goes. Ford wants to go a step further and tap the trove of financial and demographic data its finance arm holds about owners, too. "We know what people make . . . because they borrow money from us. We know if they're married. We know how long they've lived in their house because these are all on the credit applications," says CEO Jim Hackett. The message to financial markets is clear—car companies plan to cash in on the interior action.

Investors may swoon over the prospects of immersive in-car entertainment, but this twist will be a disappointing turn for policymakers. Will the cognitive surplus produced by automated commutes be eroded away by binge-watching and video games? More worrying, will the vast infrastructure of attention, which we thought was being trained on the world outside to keep us safe and sound, inevitably be turned on us inside? Instead of a new, AV-liberated proletariat—are future ex-drivers destined only to become content-farming serfs? After all, Google, Baidu, and Yandex—America, China, and Russia's biggest search engines—are among the most heavily invested in self-driving technology.

Your mileage may vary. The rich will find ways to opt out of this fishbowl future, with premium services that provide ad-free rest for the weary. But for many, getting around in the future could inevitably mean submitting to be scanned, sorted, and solicited in the salon of a self-driving vehicle.

You Are the Weakest Link

Distracted driving is defined as "anything that takes your attention away from the task of safe driving." It is an effective killer. Allow your gaze to stray from traffic for but two seconds, and the odds of a crash instantly double.

Over the years, billboards, radio dials, and fast food have all competed with the road ahead for drivers' attention. But despite the growing range of diversions on the road and in our cars, driving safety steadily improved for decades. After peaking in the 1970s, the total number of people killed each year in road crashes in the US declined rapidly, even as population and total miles driven grew. Laws requiring safer car designs, improved road layout, and better enforcement of traffic rules were crucial in securing these gains. But driver education also played a big part. People were trained to keep their eyes on the road—and for the most part, they did.

In 2005 this well-established trend suddenly reversed itself. As soon as we began to carry our phones and computers in our pockets, and into our cars, awful things started happening again. Between 2005 and 2008, deaths from distracted driving in the US jumped by 28 percent. In 2016, more than 3,500 deaths and nearly 400,000 injuries were blamed on distracted driving in the US alone.

But this is a worldwide problem. The World Health Organization (WHO) has identified distracted driving as a global health crisis, based on evidence that drivers using mobile phones are four times more likely to be in a crash. Dozens of countries, including France, Portugal, the UK, and Australia, have enacted strict laws restricting the use of mobile devices while driving. But the number of deaths on the roads continues to climb.

———

THE TOXIC INCOMPATIBILITY of high-speed car and handheld computer caught everyone by surprise. But even more unexpected is that the same bits of technology that make smartphones so distracting also hold the key to a solution—a *driverless* vehicle. Think about it. Assemble some chips, some cameras, and the cloud in one way, and put them in people's hands, and you get a public-health catastrophe. Wire them together and put them into a car, and that distraction engine is transformed into an attention machine that can replace all-too-fallible humans.

This is no easy trick. A fully automated vehicle must master three basic tasks. It needs to scan, study, and steer.

First, the computer must take in the world around it. While for humans this is an entirely visual activity, for AVs seeing involves many different swathes of the electromagnetic spectrum. Three main tools are used—radar, lidar, and digital cameras. Radar and lidar work like a bat mapping its night terrain through sonar squeaks, sending out signals at different frequencies and waiting for the echoes that reflect back from roads, traffic, buildings, and terrain. Cameras don't emit light but simply soak up photons bouncing off the stuff around them. Each of these sensors has its pros and cons. Deciding which ones to use involves a complex weighing of trade-offs among granularity (or level of detail), range, field of view, effectiveness in different lighting and weather conditions, bulk, and cost.

Radar, short for "radio detection and ranging," is cheap, rugged, and widely used. Its rapid development was one of the great technological windfalls of World War II. While radar lacks detailed resolution, its microwave beams can slice through rain and fog with ease, making it useful for establishing a proximity-detection field. Unlike the sweeping radar dishes seen on TV and in films, today's automotive radar sensors are built around a tiny "system-on-a-chip" CPU that combines many previously separate components into a single package that's easily mounted on a vehicle's exterior. Radar powers the adaptive cruise-control, parking-assist, and blind-spot-detection features in many vehicles today. It can even be bounced off the pavement and under cars in front of you to provide a view of traffic ahead.

Lidar is, in contrast, very costly, rather fragile, and still conspicuously rare. By shooting laser beams from a spinning roof-mounted housing, lidar traces out features with centimeter-scale precision. It provides a 360-degree field of view of a car's surroundings, consisting of millions of individual pixels, that's updated a dozen times per second. Lidar also has limits, however. It can't see through dense fog, rain, or snow and remains quite costly despite efforts to bring prices down. For instance, Google's self-driving spin-off, Waymo, claims to have reduced the cost of lidar by more than 90 percent over the last 10 years, from about $75,000 to less than $7,500 per AV—inexpensive enough for a self-driving taxi but still prohibitive for consumer vehicles. As solid-state designs similar to the ones pioneered for radar come to market in the coming decade, however, lidar-on-a-chip devices promise to make the technology vastly cheaper and more portable.

Despite lidar's jet-fighter sex appeal, the humble digital camera is the workhorse sensory organ of the modern AV. Cameras have huge advantages—higher resolution than radar and longer range than lidar. And when cameras are supercharged by software, nothing comes close to them in terms of value per dollar. Because they depend on ambient light, however, cameras are less effective at night.

The AV's sweep draws on a few more, nonvisual, sources. There's GPS tracks, calculated by tuning in to the beacons broadcast by satellites in high Earth orbit some 12,500 miles overhead and doing some fancy triangulation. Telemetry from engine sensors and onboard instruments rounds out the haul. All told, a fully self-driving vehicle will log some four terabytes of data each day, according to chipmaking giant Intel. That's a data smear as big as the tailings of more than 3,000 smartphone-toting citizens going about their daily business.

As step two of self-driving begins, this data is packed away somewhere warm and dry. (Not under the hood, but more likely beneath the back seat.) There's no time to ship it off to the cloud. Instead, a small supercomputer springs into action. Regardless of how old you are, if you are indeed old enough to drive, there is almost certainly more computing power in even a *partially* automated vehicle today than there was on the entire planet the day you were born. Take, for instance, Pegasus, the latest state-of-the-art onboard AV computer from chipmaker Nvidia. This modern mythical self-driving beast boasts five chips capable of performing

320 trillion operations per second. That's roughly the peak performance of IBM's Blue Gene/L, built in 2005 at a cost of some $100 million to unlock the mysteries of protein folding—and for a while, the world's fastest computer. Such is the stunning yield of high-paced improvements in computer engineering, that yesterday's wonder machines become today's household servants.

Chips like Pegasus are shifting the center of effort inside your car, from powertrain to CPU, and changing the kinds of "fuel" needed. The motor in your old car harnessed the power of internal combustion. It sucked in gasoline and transformed it into mechanical power to move you down the road. This new motor in your AV is powered by deep learning. It ingests gigabytes of data and spits out a stream of insights to guide you on your way.

Deep learning sounds more mysterious than it is. The artificial neural networks that make it work were first invented more than 70 years ago. These algorithms, loosely based on mammalian brains, were the basis of a promising early branch of AI research. But after several high-profile failures, the mainstream research community largely abandoned the approach. In the 1980s, however, a handful of scholars continued to experiment with neural networks, tackling tough pattern-recognition problems like decoding speech and reading handwritten text. By the early 1990s, the technology had advanced to the point where neural networks were put to work in banks and postal systems, deciphering billions of scribbled checks and envelopes every day.

The big breakthroughs that brought neural networks back into the limelight bore geeky names like *convolution* and *backpropagation*, a legacy of the field's long obscurity. But by making it possible to weave more than one neural network together into stacked layers (*deep*), these techniques radically improved machine learning's predictive capability. Even more remarkable was their seemingly intuitive power (*learning*). You didn't have to program a deep learning model with descriptions of exactly what to look for to, say, identify photographs of cats. All you had to do was wind the mechanism up with a million pictures of cats and it could deduce the fundamental indicators of cat-ness all by itself. This process, called "training," works by slowly calibrating the nodes within and between the stack's various layers, strengthening the connections that contribute to accurate results and pruning those that don't. Deep learning does have at least one enormous drawback, however. It is a ravenous consumer of computer power. That's why it wasn't until the mid-2010s, with the concurrent arrival of cheap and powerful CPUs and mountains of user-generated photos on the web, that the conditions were ideal for deep learning to take off.

Today's AVs put deep learning to work not to find cats but to distinguish trucks from pedestrians or tell whether the road ahead is paved with asphalt or covered in gravel. But, ironically, your AV also uses this firm new grip on reality to delude itself. That's because a driverless car doesn't so much drive itself through the real world as drive itself through a video game based on the real world. Like any good game, this one has a playing field, called an *occupancy grid*. The occupancy grid provides a geometric structure for organizing everything the computer knows about what's going on out there, including stored maps of terrain and roads and objects that have been recognized by the neural nets. The occupancy grid is the totality of your car's knowable universe.

Good games also need well-structured rules. AVs come preprogrammed to know the basic laws of physics, and traffic rules, too, but put deep learning to work again to deduce the probable behavior of turning vehicles and erratic cyclists. So-called *fleet learning* pools data across swarms of AVs to deduce, from many thousands of experiments, what to expect when navigating a left-hand turn into oncoming traffic, pulling up to a crowded curb, and negotiating other joys of modern driving. What one vehicle learns, or its mentors in the cloud distill from the herd, can instantly be instilled among all.

Finally, the driverless vehicle is ready to swing into action. With the occupancy grid occupied, the AV puts its understanding of both the natural world and human nature to work to guess what happens next. In the game board, these predictions are represented by a *cone of uncertainty* that indicates a range of possible future positions for objects based on last-observed speed, trajectory, and other factors that may influence possible course changes. The software weighs its choices and makes a decision, translating its virtual representation into the physical world. Orders are sent to switches and motors. Throttle, steering, and brakes swing the steel beast to and fro. Up to this point, everything has been guesswork, and approximation, mere bits in a register. Now, atoms are lurching about. Momentum is made manifest in mass. The rubber hits the road.

———

THIS HISTORIC MOBILIZATION of computer power is a fair response to the carnage of the motor age. The proponents of driverlessness make the case with horrifying statistics. By 2030, more people will have died at the hands of drivers every year than from HIV/AIDS, cancer, violence, or diabetes. Cars and trucks kill more children and young adults (age 5 to 29) than any other cause.

But is more technology the best answer? There are cheaper, faster, proven ways to reduce road deaths that we know work—seatbelts, pedestrian-friendly street design, mass transit, and driver education. When we dream of a *driverless* future, we abandon these time-tested solutions. We instead allow ourselves to be remade as mere apparitions in a computer game whose score is measured not by our innate worth but by the risk and liability we pose to an AV's owners.

When we embrace a *driverless* future, we also put human dignity at risk. While *self-driving* merely glorifies the marvels of meticulous machines, *driverless* defines progress as the removal of human deficiency. We celebrate the cognitive clarity of computers and chastise our own absentmindedness. *Driverless* puts us on the defensive, playing off our fear of each other, and of ourselves. It makes us, not technology, the problem. And worst of all, we surrender our rightful role as victims of the automobile age and accept full blame for its crimes instead.

Autonomists Rule

The glorification of gadgets and the devaluing of human abilities aren't enough for some AV advocates. They seek to stake out a new, sovereign territory for *autonomous* vehicles, acting independently, without external control. These ambitions have some fearing a robot revolution. But *autonomists*, as I call them, make a compelling case that we should harness machine independence to better society. There are three main planks to their party platform.

First, full autonomy is both imminent and—more importantly— inevitable. Any nascent ideology needs a rhetorical wedge to split its converts off from the herd, and followers of autonomy are united by an unwavering faith that the technology will deliver on its promises. This certainty is chalked up to the accelerating pace of improvement in the brute processing power of computers, the growing abundance of data to decipher driving environments and behaviors, and the increasing sophistication of AI techniques like deep learning.

Weaponized in the hands of pundits and prognosticators, the term *autonomous* is used to instill panicked fear of fast and devastating change ahead. To them, fully *autonomous* vehicles (which operate with complete independence) are as big a leap over merely *automated* ones (like the wire-guidance and remote-control schemes tested in the mid-twentieth century) as the motor car was over the horse. Automated vehicles are low tech and low risk, but deliver low reward. Autonomous ones are high tech, high risk, and high reward. The rhetorical split—meant to highlight the cleavage in engineering philosophies between the setting suns of the auto industry and the rising stars of Silicon Valley—is quite deliberate.

Autonomists also believe that mass adoption of autonomous vehicles will unleash a virtuous cycle of social and economic progress. If the first plank appeals mostly to geeks, this one holds a message of redemption for the masses. The automobile was once a potent symbol of the modern age, but today it is often scorned as a polluting, dangerous, and isolating means of transport. Autonomous driving, however, could turn cars into magic chariots that would never crash, and always be on call, yet somehow never be in the way. When tech titans and design visionaries raised the curtain on a shrink-wrapped vision of this too-good-to-be-true future, autonomy was the noun, verb, and conjunction that glued the whole facade together.

Finally, autonomists hold that government should play a minimal role, if any, in the driverless revolution. Truly autonomous vehicles can operate entirely within the existing road network, they argue, so they won't require costly new infrastructure. "We don't have the money to fix potholes," noted the head of Google's robot car project, Anthony Levandowski, in 2013. "Why would we invest in putting wires in the road?" What's more, markets should be left untouched. Competition would flourish, as AVs summoned by smartphone would offer consumers a universe of choices. In this, autonomists eagerly adopted ride-hail companies' push to deregulate the taxi business, which continued efforts by libertarian think tanks dating to the early 1990s and funded by billionaires Charles and David Koch.

In practice, much of what autonomists preach proves to be false. AVs simply aren't as independent as we're told. In fact, they're turning out to be the *most* infrastructure-dependent vehicles ever created! For one thing, AVs rely on the cellular grid. On stretches of road in rural America, where sizable cellular-coverage gaps exist, driverless drones would be in trouble. Wendy Ju, an AV expert at Cornell Tech, says that makers of autonomous-driving software are still trying to understand how long it is safe to operate without a cell connection—indicating that the default assumption is they cannot. AVs also need robust wireless connections to fulfill autonomists' first maxim of constant technological improvement, enabled by fleets working in concert to make sense of the world. Data from AVs in the field must be constantly shipped to superclusters in the cloud where it can be analyzed in bulk.

Now, for these transgressions of their own antidependence orthodoxy,

I'll let the autonomists off the hook. Linking up into private-sector wireless grids is, at least, consistent with their overall free-market worldview. But AVs will suckle at the public teat in surprisingly numerous and diverse ways, too. Look up to the heavens, to the constellations of satellites in orbit overhead, parked there at the expense of American, European, Russian, and Chinese taxpayers. AVs would not exist without paying homage to these celestial navigational gods every second of every day. Now, look down to the roads, without which AVs won't budge one inch. The pavement, the storm sewer that drains it, the streetlamp that lights it, the police and emergency medical teams that patrol it—all are essential functions of government that autonomists haven't yet written out of future history. Even something as superficial as lane markings, which AV cameras must be able to see clearly, will simply vanish without governments keeping them freshly painted.

Autonomists also fail to credit the crowd, that vast pool of workers mobilized to label the imagery that deep-learning systems require as input. This sea of human teachers—including, for instance, the team Google kept in India to train its first AVs and the 300,000 online gig workers of Seattle-based Mighty AI—performs endless hours of mind-numbing human intelligence tasks (*HITs* in AI jargon), the most underappreciated role in the creation story of this technology. Some of this work is done once, early in the development of AV software, to provide a baseline for algorithmic training. But many human handlers must be kept on to decipher imagery that computer vision can't interpret.

Despite autonomists' overreaches, progress toward full autonomy continues, as measured by the number of disengagements—incidents where human safety engineers are forced to intervene and take back control from stymied self-driving computers during test drives. For instance, in 2017, GM's fleet of Chevy Bolts disengaged on average once every 1,254 miles during testing on the challenging terrain of San Francisco streets, a huge leap over the previous year, when the computers balked every 235 miles.

What's unclear is whether this rate of improvement can be sustained. For companies a little further up the deep-learning ladder, the pace of advances is becoming erratic. For instance, Waymo, the industry leader, is way ahead of GM. But in 2017, Waymo managed just a 10 percent improvement in the rate of disengagement over the previous year. Then in 2018, the company made a huge leap, more than doubling the average distance between disengagements. Such herky-jerky movements forward, like a hesitant AV itself, suggest an uncertain road ahead. Will progress slow down once again, or is the industry truly closing in on technological perfection?

<hr />

TRUE AUTONOMY MAY have been a head fake all along. Even as we perfect computer-controlled driving, the meaning of *autonomous* has been stretched beyond meaning. The *Oxford English Dictionary* defines *autonomy* as "freedom from external control or influence." But in practice, as in the official definition used by the state of California, autonomous technology merely "has the capability to drive a vehicle without the active physical control or monitoring by a human operator." The only independence that's left for AVs is from us—not from other machines, networks, the law, or markets.

This slippage matters. As in the case of the automobile, AVs' symbiotic relationship with infrastructure will place a growing demand on public resources. An estimated 400,000 "fifth-generation" (5G) wireless sites will be required in the US alone, mostly to provide an umbrella of coverage for future AV fleets. The $150 billion or so that's needed to pay for it will surely come from private companies, but huge swaths of the public airwaves will be permanently roped off to shore up AVs' lifeline to the cloud. Meanwhile, the placement of 5G networks' dense mesh of antennas has already triggered a backlash from municipalities across the US.

It's unlikely that the bait and switch will end there. Having failed to bootstrap the driverless revolution, the ideology of autonomy will soon acknowledge its new reality. Libertarian trappings will be stripped away, and a more pragmatic neoliberalism will creep in. Rather than push government away, AV makers will pull the public sector deep into the project. In the long-gone days when it seemed that the auto industry and government would have to team up to lay guide wires across the land, car companies like GM embraced a vision of joint "national purpose" for the needed improvements. It seems inevitable that similar rhetoric will be mobilized to recruit taxpayers to pick up the bill for paving the ghost road.

A New Lingo

The spinmeisters of the driverless revolution have their work cut out for them. Our fear of intelligent automobiles is as deep-rooted as our yearning for them. Remember Christine, the haunted homicidal 1958 Plymouth Fury brought to life by Steven King in the 1983 novel and film? Who could forget that first sinister glimmer of vengeance in Christine's headlights as Arnie's drifting attentions to young Leigh take center stage?

We're right to fear these machines. Much of the thinking behind these technologies goes directly against closely held values, ignores social norms, and threatens our livelihoods. But our fear works in their favor too. By turning us against each other, and ourselves, this fear becomes a tool to sell the technologies of tomorrow. By stoking us to such a fever pitch with the rhetoric of human inferiority, obsolescence, and replacement, the wordsmiths of the self-driving, driverless, and autonomous future paralyze us. They make us easy pickings for "disruption"—that profitable, chaotic, and painful process of change that often accompanies new technologies.

But we may get the last laugh. Because there's a case to be made that automation will make driving *more* important, not less. And it will put people in ever-more important and exciting roles—not push them to the margins.

For starters, for the foreseeable future, making and selling technolo-

43

gies will be much easier if they *enhance* drivers' abilities than if they *replace* them. That's not just because of the engineering challenges involved; it's also because of the reasons people want automation, which often have little to do with saving labor. In the 1950s, when electric appliances were widely introduced in American homes, anthropologists who studied domestic life noticed something peculiar. Women using these supposedly time-saving technologies weren't spending less time cleaning. But there was a big change—the socially acceptable standard of cleanliness rose. Similarly, it's easy to imagine automation simply raising our standards of safe driving, fast driving, or green driving in the future. Supercomputers won't fully take control but instead will serve as wise and willing life coaches for superattentive soccer moms, hypermiling granddads, and hot-rodding hunks. Automation won't be a binary thing that's on or off, and its value won't be measured only in time. We'll use it to invent all kinds of new software-assisted ways to drive that match our diverse needs and desires.

Next, drivers do more than just drive! This is such an obvious insight—inescapably so when we talk about professionals at the controls of taxis, trucks, and buses—that it's dumbfounding how little this has factored into debates about AVs. Professional drivers load and unload cargo, refuel, prepare and serve food, remove waste, perform routine maintenance and emergency repairs, handle security, and treat medical emergencies. Yet the amount of research and development being done to automate these functions is negligible. While some can certainly be automated in whole or in part, it will be more costly or impractical to do so.

What's more, as they have done with every other automotive innovation, early adopters will poke, prod, and hack AV technologies for high performance and for delight—because they're people who *like driving*. Tesla transformed the electric vehicle, from the motor vehicle's crippled cousin to its superior successor, by exploiting the instantaneous torque electric motors create to deliver astonishing acceleration. Already, German luxury automakers see opportunities to "tweak the performance of electric engines to give a premium experience," oxymoronically betting that "passengers will still pay extra for a better driving experience even if they are no longer at the wheel." Can we expect more exotic hybrids of human-computer driving that supercharge our experience behind the wheel—or replace the steering wheel with something new and remarkable entirely? Imagine—rather than a world of machine-managed pod cars—highways filled with mind-controlled motorcycles, toddlers at the wheels of tractor-trailers, and lane-weaving gearheads gunning their rides to superhuman extremes. Software could make it all possible. Market mechanics and our own mores may make these wild ideas likely.

Finally, the delegation of driving to computers may make operating a vehicle *more complex*, not less. This has long been the case in aviation. Pilots of today's aircraft now require much more training. "Technology does not eliminate error, but it changes the nature of errors that are made, and it introduces new kinds of errors," argues Captain Chesley Sullenberger, the US Airways pilot who safely landed an airliner in the Hudson River in 2009. The further the spread of automation in the cockpit, the thicker the machines' manuals have become.

———

OUR PRESENT TERMINOLOGY for the journey ahead is deeply problematic. The proof is in the public attitude. According to annual surveys by the American Automobile Association, a motorists' advocacy group, in 2017, 63 percent of US drivers reported they "would be afraid to ride in a fully self-driving vehicle." Six months later, after several highly publicized crashes of partially automated AVs, the rate of skepticism had increased to 73 percent. Halfhearted attempts over the last decade by automotive engineers to classify vehicle automation into six levels have only confounded the public. Does anyone besides engineers begin numbered lists with zero?

There's a growing sense in the AV industry that the technology's success will depend on finding better words to describe it. For instance, in 2019 Waymo announced a new unattended "rider-only" taxi service that left human safety supervisors behind. Thankfully, the basic requirements for this new lexicon are much clearer than they were just a few years ago. At a minimum we'll need phrases to fill in the uncanny valley of partial automation—names for the ambiguous situations arising from divided control and from the misunderstandings and miscommunications that arise from subtle differences in human and machine perception.

But there's also a whole vocabulary to be crafted to explain the unfamiliar space-time of teleoperation. Today Waymo's "fleet response" team members monitor vehicles in the field from a central command center in Chandler, Arizona, and "weigh in with an extra set of eyes" whenever an AV meets a surprise on the road. These operators don't steer but approve AV-planned detours like driving around a double-parked vehicle blocking a one-lane street. The trucking industry is also experimenting with remote human control. Starsky Robotics, a startup, is developing big rigs that drive themselves on the highway but when traveling on local streets are steered by human operators in a distant control center. What do we call this scheme, which resembles the way harbor pilots take control of big ships as they approach the coast? Similarly, but at a much smaller scale, MIT's Sangbae Kim builds robots that move over terrain under AI-guided locomotion, while remote pairs of human hands take control of arms and cranes during fine-grained manipulations of the world in front of them. None of the words we use now—*self-driving, driverless,* or *autonomous*—captures what's important or distinctive about these human-machine hybrids. But they'll need good labels if *we* are to understand what to expect of *them* out there.

Priority number one, however, is dismantling the language of AVs today. We need terminology that shifts the focus away from computers and cars and back to us. Much as with other quickly abandoned terms for new technologies—like *horseless* carriage and *cellular* telephone—*driverless* car and *autonomous* vehicle are engineers' terms that define the new in opposition to the old. They obscure the practical nature of the invention in the details of its inner workings and theory of operation. In contrast, *automobile* and *mobile phone* stayed with us because they explained the benefits of new technology to everyday people. We'll need words that, like these, tell us a story about what's to come and why it matters—not about what we're leaving behind.

This isn't just semantic nitpicking. Words will be the key that allows us to see ourselves in this future. They must remind us, every day, that we are the masters of the technology of tomorrow.

3 The Origin of (Vehicular) Species

We might need self-driving buildings as well as self-driving cars.

—Astro Teller, Captain of Moonshots, X

Modern programming languages distinguish between two kinds of "objects," or bundles of code and data—those that are *mutable* and those that are *immutable*. Mutable objects can change; immutable ones cannot. The scientists and engineers who design these protocols are such sticklers because computers are by nature so dynamic. Their very name is a verb for mathematically changing one thing into another (*computer, compute*). So if you want some information to remain unchanged for any meaningful bit of time, you need to fence it off and put up a sign.

Human beings couldn't be more different. We prefer things to remain as they are. We are, in fact, so comfortable with the absence of change that in the 1980s behavioral economists coined a term for this tendency—*status quo bias*. Give us a choice between, say, different investments for our retirement funds, and we'll make up our own mind. But give us that same choice and tick one option as the default—and we're highly likely to go with the flow.

One reason why we abhor change is fear, a related phenomenon behavioral economists call *loss aversion*. Sometime long ago in our evolutionary history, it seems that natural selection favored those individuals inclined to weigh the risk of future losses more heavily than the possibility of future gains. Loss aversion is what keeps us from pursuing our dreams, confessing a crush, or trying a new flavor. We simply find it hard to imagine the world getting any better—or more precisely, we can't imagine it being any different—than it is now. So we stay put.

These predispositions toward continuity are on full display as the age of AVs dawns. Day after day, digital disruptions are changing everything about how we live, work, and play. The future is constantly in flux. Yet our driverless daydreams stubbornly default to a familiar form, the family car. They hardly differ from the visions that tantalized our grandparents generations ago.

Imagine it's Christmas 1956, and young Grandma cracks open the latest issue of *Newsweek* to find an intriguing pitch. A full-page illustration depicts a well-coiffed family of four, seated in a glass-domed sedan, gathered around a game of dominoes while they cruise. "One day your car may speed along an electric super-highway, its speed and steering automatically controlled by electronic devices embedded in the road," the caption explains (Figure 3-1). It's a scene of purely modern domestic bliss, this prototypical nuclear family rocketing across the open range. But the world wasn't yet ready for the prospect of *computer* control. This scene is animated by a more familiar phantom force. "Highways will be made safe—by electricity!" Bold promises delivered through automation, by now familiar to you, are in store. "No traffic jams . . . no collisions . . . no driver fatigue."

In recent years this old ad has enjoyed a renaissance—in blog posts, TED talks, and startup pitch decks. It's a favorite of those looking to speed self-driving technology along. Yet for all the nostalgia it evokes, this is a picture of a future that never existed and never will. In this imaginary

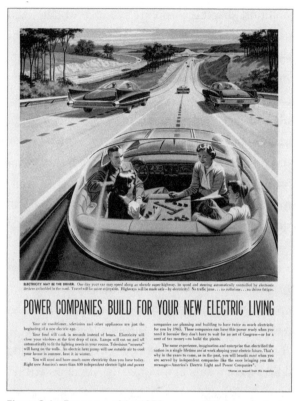

Figure 3-1. *Future car of the 1950s.* An electric power industry promotional advertisement in *Newsweek* magazine in 1956 promotes a future of self-driving cars to millions of readers.

THE ADVERTISING ARCHIVES/ALAMY STOCK PHOTO.

world of tomorrow there are no trucks—and no commerce of any kind, for that matter. Yet today more goods move by road than ever, and the biggest surge is still to come. Here we're shown that families are composed of a man, a woman, and children, who travel everywhere by private car. Yet such families are now a minority in the US, and still fewer get around this way—except perhaps on our ever-more-infrequent holidays. And this future features no buildings, yet in our world people live in cities in unprecedented numbers.

But what fascination this myth holds. Its mistaken assumptions echo across generations, shaping corporate product strategy and consumer preferences to this day. If you replaced America's Electric Light and Power Companies—the lobby of electric power utilities that sponsored this ad—with Tesla or GM, you could run the same spread today. No one would be the wiser.

Yet while this picture got everything wrong about the messy reality of the future—and still does—we never question its most important assumption. Why, in this coming world of wonder, are we still getting around in *cars*?

Even in our wildest aspirations, it seems we can't free ourselves from the status quo.

WE WEREN'T ALWAYS so short on imagination. Soon after the automobile arrived, we got busy stretching out, tucking in, lightening down, and beefing up. Horseless carriages quickly evolved into vehicles with a stunning variety of shapes and sizes, and we never stopped tweaking. There are sedans and SUVs, compacts and minivans. These get turned into taxis and limousines, police cars, and hearses. Trucks are made for hauling stuff instead of people, and come in an even larger range of shapes and sizes to match their diverse cargoes—box vans, flatbeds, tankers, and pickups, to name but a few. Some get special mods to do the dirty and dangerous work of city upkeep—there are street sweepers, steamrollers, garbage trucks, and snowplows. Darting in and out among the biggest beasts are the two-wheel types—scooters, motorcycles, and mopeds. And don't forget the buses, elephantine people-haulers of the urban jungle.

It's pure delight to point out and name all the things people drive—a popular and timeless theme for children's books. Yet even though we learn this truth from an early age, our brightest futurists forget it the first moment they slide into an AV, where they prefer instead to imagine a world of computerized order and uniformity. We've already seen how driving won't be a binary choice, when computers don't so much take over as become our copilots in an automated future. In the same way, automation won't *reduce* the variety of vehicles we use, but will instead radically *expand* it.

This vehicular variety is the next thread in the first big story of the driverless revolution, *specialization*. For more than a century, our vehicles have been variations on a theme—the familiar chassis of the horseless carriage, distorted beyond recognition but never abandoned. This new age of automotive invention, however, will push that envelope until it tears. Driverless vehicles will grow to the size of small buildings, so big that they can move only across the open countryside or through empty city streets at night. Or they may shrink to little more than enclosures for the tiny electric motors that make them go. How about self-driving shoes? Punch in an address on your mobile, press Go, and enjoy the ride while flipping through your inbox. It's much closer than you may think.

Computer control will also offer new possibilities to build vehicles that are much faster, or much slower, than automobiles today. Zero-occupancy vehicles that don't need to keep passengers comfortable will be overclocked like high-powered gaming PCs—taking turns at high-Gs and throwing in evasive maneuvers that would shatter bones. But they'll also throttle back to a snail's pace when needed to dial down danger, noise, wear and tear, and fuel consumption. All without the risk of boring a soul.

The point is this—future AV designers won't just think outside the box. They will drive right over the box and do algorithmically optimized doughnuts on it. Code will let them make the most of the endless hours and open expanses of the ghost road.

———

THE SEEDS FOR this blossoming of vehicular variety were planted in the recent past.

The most important shift has been in our daily travel patterns. It used to be that every commuter, student, and homemaker made the same daily journey—home to work, home to school, or home to market. Stops along the way were rare. These unchanging patterns concentrated trips along predictable routes with clocklike regularity. But now, our movements are more individualized. We stick less to traditional work and household schedules, and take more trips that have nothing to do with jobs or families. And we "trip chain," as researchers call it, making multiple stops instead of driving directly to and from work, "to buy coffee in the morning, to drop off and pick up children at day care or school, to visit the gym, and to buy groceries." As a result, we visit more places every day, for shorter periods of time, over a greater range of territory than ever before.

Second, the way we make vehicles has changed. For a half century, a handful of giant companies dominated global auto production. But new methods of manufacturing, like 3-D printing, mean that smaller firms can now profitably produce high-quality vehicles. Meanwhile, an increasing portion of the added value in vehicles comes from software controlled by companies outside the traditional automotive industry. This means firms like Google and Baidu can muscle in, while traditional carmakers are relegated to the role of commodity steel-stampers. Together, these changes create a paradox. The biggest factories today are the largest humankind has ever built (and they make phones, not cars!). But a vast sea of little plants can ship more kinds of specialized vehicles than ever before.

Third, electrification and automation are symbiotic technologies that work best in combination. Electrification paves the way for automation by extending computer control throughout a vehicle's entire steering and propulsion system, creating points for software-based innovation. But automated driving is also a boon to electric vehicles, by making it easy to coordinate recharging schedules and locations, which spreads the strain on power grids and allows heavier use of renewable generation. And in the more distant future, when autonomy is widespread and collisions have become a thing of the past, vehicles will no longer need to withstand the force of high speed collisions. Lighter frames could be used, reducing vehicle weight by more than 20 percent. With less mass to push, an AV's electric motors and batteries could then be downsized too.

These shifts may herald a realignment of automaking activity away from traditional hubs to high-tech regions. Already, big car companies are struggling to adapt. But whether the future is made by a handful of companies or hundreds, the pattern is clear. The vehicle market is fragmenting and automation will further accelerate this process. It is easier than ever for upstarts to move in, and a panoply of new technological synergies await exploitation. And it's this turmoil that will give birth to the new types of vehicles that will shape our world for decades to come.

Starships and Shuttles

I first stumbled across the budding diversity of the driverless future-to-be in the autumn of 2016. I was in Tampa for the Florida Automated Vehicles Summit. For the fourth year in a row, this event had gathered government officials and industry leaders from across the Sunshine State to talk about the prospects for driverless technology. With an economy that lives or dies on its future appeal to footloose residents, retirees, and resort visitors, maintaining a high-quality transportation system has long been a priority for the state's growth machine.

As I pushed my way through the gathering crowd in the hotel lobby, I caught a glimpse of a goofy-looking kid grinning by the bar. His face was backlit by the glow of a tablet computer, and he was surrounded by a pack of old guys in their best business casual, scotches in hand. So far, the usual conference crowd.

I made a beeline for the buffet, but I didn't get far before R2-D2—or what could have been his long-lost cousin—swept in to block my way. The six-wheeled buggy jerked to a stop, its stubby antenna quivering for a moment. The kid tapped his screen and R2's top popped open, revealing a cooler of ice-cold beer and soda. I bent down to get a look at the diminutive delivery wagon's license plate. STARSHIP, it read, in bold black letters.

These craft are easy to make fun of, but they intrigue me all the same. As I soon learned, the management consultants (the guys with the scotches over at the bar) call this whole emerging category of driverless droids "automated ground vehicles"—or just AGVs for short, hitting the *G* hard to distinguish them from their larger cousins. I call them *conveyors* (Figure 3-2, page 61) instead, because they remind me of buckets on a warehouse belt as they jog along the sidewalk. This one hails from Estonia and was built by Starship Technologies, a startup founded by the same guys who created Skype. For over a year, the company had been testing the little droids to deliver lunch to college students in Tallinn, the capital. Their thinking was that European cities could adopt conveyors as substitutes for the fleets of delivery cars used by fast-growing companies like Postmates and Deliveroo, which were adding to traffic, noise, and air pollution on neighborhood streets. Eventually, they thought, the whole world would catch on to the idea.

It was hard to see this contraption living up to its creators' android ambitions. To me, R2 looked hopelessly flimsy, and more annoying than useful. I tried to picture thousands more rolling through neighborhoods, endlessly carting stuff from depot to front door. Peering in for a closer look, I started to imagine all the things I'd order up via automatic bucket brigade. Milk, eggs, toilet paper . . . all those needs that send you out into the rain on emergency errands.

But why stop there? The machines built by Marble, another conveyor maker, are less sleek than Starship's but a helluva lot bigger (about the size of a 1980s photocopier). They carry "up to four bags of groceries, six shoe boxes and 10 hot meals," according to one report. Screw the Sunday shopping! I'll just get some sun on the back porch and send my robot to the supermarket instead.

Conveyors illustrate all of the forces driving vehicular variety in the driverless revolution. They are highly specialized, dispensing with the thousands of pounds and hundreds of cubic feet of chassis and motor it takes today to deliver small items by car or van. The tiny Starship's construction employs a global web of design and engineering, manufacturing, and field-testing that spans Europe, Asia, and North America—an all-but-unthinkable reach for a firm that's raised a modest $42 million and employs fewer than 250 people. The conveyor's electric, autonomous drive makes it small and silent, perfectly compatible for residential areas and retail districts.

That stealth comes at a price. Conveyors aren't roadworthy in the conventional dog-cat-dog sense of mass and momentum. (Starship vs. SUV, SUV wins. Starship vs. subcompact . . . subcompact wins.) By design, most can operate only on sidewalks and crosswalks, which puts them in direct conflict with pedestrians, pets, and wheelchair users (Starship vs. pooch, Starship wins). When a handful of startups began testing conveyors in San Francisco in 2017, a sharp backlash ensued, spurring the city's Board of Supervisors to enact a temporary ban.

Yet while such fears aren't misplaced—it's easy to see pedestrian areas becoming overwhelmed by conveyors—it may be possible to program them to be so deferential that we may not mind. That's the thinking behind an effort at MIT's Aerospace Controls Lab, which is using deep learning to help computers understand the "intricate sidewalk ballet" described by urbanist Jane Jacobs in her 1961 classic *The Death and Life of Great American Cities*. Researchers have already successfully programmed a conveyor with enough sense to learn our unwritten rules of body language and smoothly move through crowds. But if faking human moves doesn't solve sidewalk gridlock, conveyor makers may simply try to make their machines more cute. At the University of California, Berkeley, Kiwibot's conveyors serve up snacks to students and "have been adopted by the community," according to a company official. The droids show up on social media around campus and have inspired student Halloween costumes. When a bad battery set one ablaze, "students held a candlelight vigil for it."

Conveyors aren't the only AVs purpose-built for drayage, the traditional term for hauling goods over local streets. Two Silicon Valley companies, Nuro and Udelv, are building van-sized variants called *mules*

(Figure 3-2). Both companies were started by alumni of (respectively) Google's and Tesla's self-driving teams, and both launched in early 2018, capping a sudden surge of venture-capital investment into this AV category. These larger bots, with their heavier hauling ability, are attracting attention from carmakers, too. Ford plans to launch one based on a pickup truck chassis in 2021—the struggling automaker's F-150 line accounts for one-third of the company's production but 90 percent of profits. And a Mercedes-Benz Vans concept vehicle that debuted in 2018, called Urbanetic, "eliminates the separation between people moving and goods transport." Fitted with either of two interchangeable bodies atop a common motorized sled, the 5.14-meter-long vehicle can carry 12 passengers or several tons of cargo.

THE DAY AFTER my close encounter with the conveyor, I gained another vista on the driverless future from a perch on Tampa's historic streetcar, a replica of old lines that used to crisscross the city and now made its way past my waterfront hotel. Since opening in 2002, the TECO Line streetcar has played a pivotal role in tying together the disparate pieces of Tampa's downtown renaissance. The 2.7-mile route snakes its way up Old Water Street, taking a leisurely 15 minutes to reach Ybor City, a historic immigrant district. With a top speed in the teens, it produces just enough of a breeze to compensate for the oppressive Florida humidity. Ybor, once known for its cigar factories, now enjoys a new life as the city's arts and nightlife hub. There's so little of this un-air-conditioned, pre-geriatric Old Florida left, you have to grab it and hold on to it when you find it. But after a long stroll and a quick puff on a Cuban, it's back down the line, toward home.

While the old trolleys looked and felt like a relic from days past, they did put one of the AVs on show at the summit to the test. As we reached my point of origin, there was a new AV waiting outside the hotel. Imagine a gondola straight out of the Alps but with four wheels on the bottom, and you'll start to get the picture (Figure 4-1c, page 104). A cheerful young woman in a sharp suit beckoned me to board.

To describe what happened next I'd have to take you back in time, sit you next to me, and ask you to be still for a moment. I'd ask you to close your eyes and try to guess the exact moment when, with a silent and imperceptible shove, we set off. You'd get it wrong, and I'd explain that unlike internal combustion engines, electric motors lack that slip-and-grab feel that even the smoothest transmission delivers as the motor and shaft connect. Electric motors just go, first slowly and then faster and faster. You'd say how pleasant it felt, and I'd say that I'd often wondered if our children's children would think of our cars the way we thought of black-and-white TV and rotary phones. We'd both nod and sigh.

At least that's how I imagine it would happen. The real experience of riding in a *driverless shuttle* (Figure 3-2), as this new class of self-driving people-mover is called, was somewhat more baffling. To start with, the future's top speed was slower than I expected. We'd barely begun to move when the EasyMile EZ10, a French-made buggy, stopped accelerating. Designed to operate as a connector over short distances, often in pedestrianized environments, a driverless shuttle puts a premium on safety and efficiency rather than speed. Outside, I could see the streetcar coming around the bend, on another of its endless loops up and down the line. More than a century of progress separated these two transportation technologies. But as we were easily overtaken, I felt left behind.

Disappointed, I sat down and reset my expectations. With two benches seating four across and facing each other across a small standing area, it was cozy inside but not claustrophobic—roomier than a car, but more intimate than a bus. Conversations bubbled up, people pointed out the windows as the city scrolled by. There was even enough room to get up and mingle. My mood brightened, and my thoughts turned to the well-worn literature on "third spaces" in cities, those informal gathering places that nurture real-life social networks. The vibe felt just right.

Our jaunt was shorter than the trolley ride, a half mile up the quay, demonstrating the shuttle's use for covering what public-transit wonks call *the first mile* or (alternately) *the last mile*. It's these transit-starved gaps between main-line stations and the front doors of office buildings, schools, hospitals, and homes that create such high hurdles for people otherwise inclined to take transit. I imagined the whole thing powered by renewable sources like solar and wind, or (as in France) nuclear power, and my excitement built.

On the return trip, my attention turned outward. With their wrap-around windows and raised coach floor, driverless shuttles provide a perspective on the street akin to what nobles might have experienced traveling by open-air carriage in years past. Unlike the subway, where you're cut off in a high-speed bubble, on the surface the theater of the street surrounds you. Silent, gliding . . . far more than any experience in a self-driving car, this may be as close to a mythical magic carpet ride as we're likely to get.

The trip ended, and as I disembarked, yet another AV made its debut— a Tesla S. As I stepped off the ungainly shuttle toward the sleek, sporty bit of Silicon Valley steel, I realized I'd passed between one possible future and another—from car-lite commune to self-driving suburb. This California roadrunner was everything the French snail wasn't—sleek, sexy, free-range, and fast. And the DNA of the two companies couldn't be more different, pitting Silicon Valley's venture capital–fueled investors against the industrial giants of the French heartland—EasyMile's chief financial backer is Alstom, a maker of trains.

Figure 3-2. *Conveyors, mules, driverless shuttles, and taxibots.* (a) Light-capacity conveyor with insulated compartment for on-call delivery up to 2 km. (b) Medium-capacity conveyor with refrigerated compartment for delivery of perishables up to 5 km. (c) Light-duty mule for street-corner parcel pickup, returns, and automated vending. (d) Heavy-duty mule for drayage between fulfillment centers and curbside unloading zones. (e) Driverless shuttle for last mile, up to six passengers seated and two standing. (f) Taxibot based on current production minivan, seats up to six passengers. (g) Purpose-built wheelchair-accessible taxibot, seats four passengers and attendant.   DASH MARSHALL.

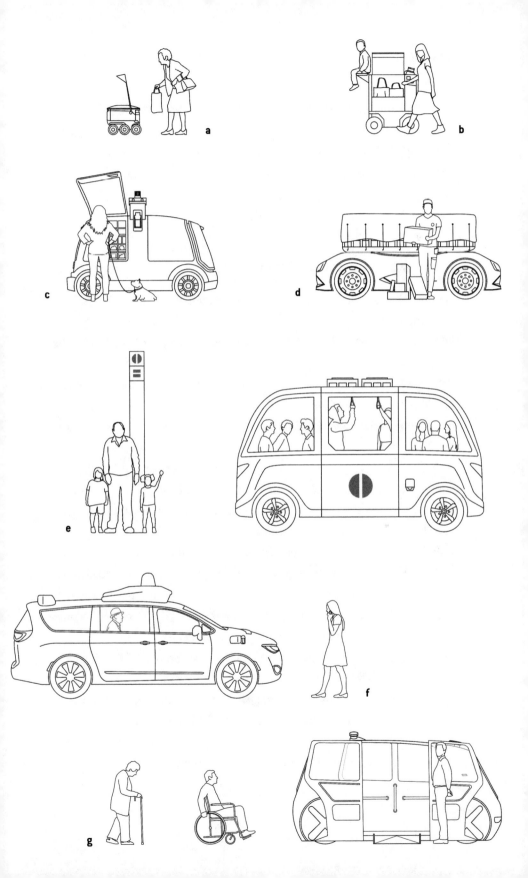

a

b

c

d

e

f

g

Which of these is the better bet to be the Model T of the twenty-first-century city? The Tesla is merely a better car. But the driverless shuttle represents a wholly different kind of luxury. Yes, this self-driving ski gondola is clunky. But it's only the first attempt to tailor a made-to-measure AV to the shape of sustainable, equitable twenty-first-century communities. The fit isn't quite perfect, but it's a very promising start.

Rovers and Software Trains

I confess. I was fooled on April 1, 2016, by a bunch of Dutch engineers. That morning, even before I'd taken my first sip of coffee, or checked the calendar, I scrolled through Twitter on my way to the bathroom. A YouTube link, the Google logo, and a thumbnail beckoned. "Introducing the self-driving bicycle in the Netherlands," read the caption. I clicked, of course. I let about a minute of the farce play out before I checked the date.

Maybe I just *wanted* to believe. The idea of a self-driving bicycle isn't as silly as it might first seem. As a teen, I used to get a kick out of "ghost riding"—leaping from my 10-speed bike to send it cruising off on a gyro-stabilized journey down the street. My friends and I would play a game to see who could get theirs to go the farthest. The secret was all in the dismount. The slightest shearing force on the seat as you pushed off and you'd send your Schwinn careening off-centerline. So when self-driving cars started to show their mettle, I wondered now and again whether bikes would be next. In 2001, after the unveiling of Ginger, Dean Kamen's infamous Segway self-balancing scooter, I hoped that our imaginary ghost cycles might become a reality. But I didn't dare tell anyone about my little daydream.

I wasn't the only one to imagine such things. But it would take a few more years to strip down the self-driving car's ambitions to cycles. Months before my April Fools' humiliation, I learned about an AV called the "persuasive electric vehicle." The brainchild of MIT Media Lab researcher Ryan Chin, the bike was a gear geek's wet dream, mashing up every quirky idea in bicycle design. It had three wheels, a recumbent riding position, electric pedal assist, a cargo compartment, and self-driving software. It was like a DIY project ripped from the pages of *Make* magazine, but Chin and crew were serious. They'd already cut a deal with the diminutive nation-state of Andorra to test the tiny AV's suitability for shared passenger and parcel service.

While Chin's tricycle confirmed my hunch about the potential for putting self-driving tech into bikes, it was the link to bike sharing that intrigued me. Bike-share is a dead-simple idea that has revolutionized how we ride in cities. Today, millions of shared bikes serve a thousand cities worldwide. They provide 24-7 local transport that's faster than walking, is more pleasant than transit, and doesn't emit a trace of greenhouse gas in the process. And as clever as Google's pranksters were when they poked fun at the idea of a self-driving cycle, they overlooked how useful the capability for riderless operation could be in making bike-share systems work better.

As it turns out, the bike-sharing revolution began in Amsterdam, the setting for Google's spoof, way back when Stanford's scientists were still cooking up their first self-driving robots. In 1965, the members of the countercultural collective Provo launched an experiment to push back against the creeping domination of the automobile by placing several painted "white bikes," as they were called, onto city streets. The idea was an instant hit with the hippie class. There's even an old photo of John and Yoko with one. But the project's antiestablishment roots were deep. Despite several attempts it never secured long-term support from Amsterdam's city government.

After the white bikes, bike sharing vanished until a new generation of activists in Copenhagen revived the idea in the 1990s and succeeded in getting buy-in from authorities. Then, Paris's Vélib system opened for business in 2007, immediately becoming the photogenic ambassador for the idea's viral global spread. (Ironically, because Amsterdam is already so overrun with privately owned bikes, city authorities have stubbornly resisted bike-share.)

What changed between the 1960s and today to make bike-share a sudden success? Smartphones were one key missing ingredient. Much as they powered the rise of ride-hail, smartphone apps are the tool that makes it easy for us to locate available bikes, unlock them, and pay as we go. In a similar fashion, low-power mobile computing is also crucial to bike-share's light-footprint infrastructure, which allows solar-powered, wirelessly linked stations to be slipped into the underused interstices of cities with minimal cost and fuss. "Docks," as they're called, can be installed in spaces no one else wants—along curbs, in corners of plazas, and outside the entrances of transit stations. New York City's Citi Bike docks, for instance, are delivered by flatbed truck and switched on without laying any wires, cables, or pipes. They aren't even bolted to the street surface.

As lean and mean as today's bike-share systems are, it wasn't long before we ditched the docks, bringing us back to the original 1960s setup. *Dockless* bike-share does away with fixed infrastructure by taking mass-produced commuter cruisers and logging them into the internet. Without the need to secure spots for stations, or pay for their construction, dockless systems expanded the footprint for sharing into new territory on a previously unimagined scale. Most notably in China, a perfect storm of factors—fast-growing cities, cash-flush investors, and cheap manufacturing—unleashed a wave of startups. In the first six months of 2017 alone, the number of dockless bikes in Shanghai tripled, from 500,000 to 1.5 million. Thickets of hastily parked bikes quickly piled up at bus and train stops—a kaleidoscopic urban sculpture made all the more brilliant by each brand's distinctive color of the rainbow.

The bane of any shared-vehicle system is *rebalancing*. Since rides are concentrated along common daily travel routes, it doesn't take long for vehicles to pile up at popular destinations. As a result, by the middle of rush hour, idle bikes accumulate in downtowns, at schools, and around transit stops, while in outlying residential neighborhoods there are no more wheels to be had. Sometimes rebalancing happens organically. A bike-share operator can wait until the evening rush and let people do it themselves by ferrying bikes back to their morning starting points. But during those long layover hours, your vehicles lie idle while your customers are at work. Most sharing systems try to manually rebalance at least once a day. But this restocking is costly and carbon intensive, because it's done by loading bikes into vans and trucks and carting them across town.

Rebalancing in dockless systems is also mind-numbingly complex, because riders drop vehicles at many different locations. Nowhere was this more clear than in the astonishing spread of shared electric scooters in 2018. Charging one dollar to hop on, plus 10 to 15 cents per minute, e-scooter services like the one offered by Venice, California–based Bird Rides offered the fastest, cheapest, most convenient, and most fun way to get around. But the huge number of vehicles and the speed with which Bird deployed them meant finding a new approach to staffing up teams to rebalance the wayward flock. What's more, scooter companies faced a severe backlash from communities, which objected to the haphazard deposits of these diminutive personal vehicles on curbs, on street corners, and everywhere in between.

Bird's answer to rebalancing was as comical as it was practical—hire a gig-worker army of "bird hunters" to round up, recharge, and redeploy scooters for a bounty of five dollars per vehicle. The scheme seemed too crazy to work—until it did. And with its decentralized approach to rebalancing, Bird's scooter network expanded even more rapidly than dockless bike-share systems had the previous summer. There was a big problem though. Bird soon found itself with the same challenge confronting ride-hail operators Uber and Lyft. Workers were taking home all the profits.

Here's where automation comes in. The Google gang's facetious scenario incorrectly assumed that the killer app for self-driving bikes was hauling lazy people around. The shift to dockless revealed something altogether. Automation was the key to corralling thousands of wayward rides on the cheap. "The pain point for scooter operators is to better maintain the scooters at a lower cost," says Gao Lufeng, the CEO of Beijing-based Segway-Ninebot, summing up the equation succinctly. In

2019, the Chinese company launched the world's first mass-produced self-driving scooter. Even at a price of nearly 10,000 yuan (about $1,400), some 10 times the company's human-steered models, this miniature AV is expected to offset its high initial cost by slashing the labor involved in rounding up rides.

Scooters are just the beginning, however. Personal electric runabouts are one of the hottest areas of innovation today, streaming onto our streets in growing numbers and in all shapes and sizes. There's Onewheel, a skateboard that balances itself on one giant, rubber, software-controlled, electric-powered wheel. Riding it can be described only as something like surfing the street. Or Superpedestrian, the MIT Senseable City Lab spin-off whose smart wheel turns any bike into an electric-assist one. Vehicles like these are but one software update away from autonomy.

I call this class of personal, electric, autonomous, and always-ready AVs *rovers* (Figure 3-3), and I expect them to swarm our streets soon.

They'll certainly be far easier to invent than self-driving cars, since their self-driving software will be as stripped down as the rides themselves. That's because they can avoid larger vehicles by taking sidewalks and side streets. And since they weigh so little, they won't do much damage when they crash. What's more, since they'll eliminate the cost of rebalancing, rover companies can grow even faster and further than scooter outfits did.

Rovers will radically change how we get around neighborhoods. Instead of your having to hunt for one down a strange street on a dark night, they'll come *to you* at the press of a button. And while dockless bikes and scooters pile up at popular drop-off points, self-driving ones will simply tuck themselves away in back alleys, empty parking garages, or underneath highway overpasses to recharge.

Like Bird and its ilk, rovers will initially appeal to the young, affluent, and tech-savvy. Unsurprisingly, the group most eager to zip about on these little magic carpets is people under 35, who find these machines to be a perfect match. But it's a mistake to think rovers aren't for everyone. Automated driving and electric motors together will provide a kick and a helping hand for those too old, too young, or medically unable to fully pedal, push, or steer under their own power—expanding the market for bike-share and scooter-share far beyond their current base. What's more, the elderly and nonambulatory will be huge markets in their own right—the number of Americans in wheelchairs grew by half between 2010 and 2014, and could pass 12 million by 2022. Hitachi, a big maker of electric wheelchairs, has developed AV prototypes of its flagship models. Segway's S-Pod, a self-balancing two-wheeled chair, will have a stunning top speed of 24 mph.

Rovers' broad potential to produce the first breakout success of the driverless revolution may be what's drawing the attention of the big guys, too. Uber began recruiting engineers for a new group, Micromobility Robotics, in early 2019. Facebook has already patented a self-balancing electric motorcycle. But cornering the market on rovers tomorrow won't guarantee success forever.

What may surprise us most about rovers is their rapid replacement cycle. They are, in a very real sense, disposable vehicles. One analyst estimates that Bird recovers the up-front cost of its scooters within four to seven weeks of putting them on the streets. That allows the company to constantly replace battered vehicles with new and better ones. But in 2019, the company launched the Bird Two, a more durable scooter model that carries built-in maintenance sensors to summon help when service is needed.

With such speedy swaps will come fast-changing fashions. Product cycles that automakers and conventional fleet operators measure in years and decades . . . the rover business will measure in weeks and months. Lime, a dockless operator, put no fewer than nine versions of its flagship bike into service during its first year and a half of operation, a stunning pace of improvement. VeoRide, a dockless bike-share company, claims it can turn a new design idea into on-street hardware in 15 days.

THE LARGEST AVS won't evolve as fast as the smallest, but they'll bring exciting innovations, too. That much was already clear in 2016, thanks to one of the most dramatic AV demonstrations to date. On a 120-mile stretch of Interstate 25 in Colorado, a computer-controlled tractor-trailer hauled 50,000 cans of beer to make what's claimed to be the first-ever commercial delivery under self-driving operation. The self-driving tractor-trailer featured a liquid-cooled supercomputer that packed more power per ounce than any before.

"It's like a train on software rails," said Anthony Levandowski, the founder of Otto, the startup responsible for this technological feat. At first, it seemed like an odd way to describe the achievement. After all, what makes trucks so useful is that they don't need rails. That flexibility, to go anywhere at any time on an extensive network of roads, allowed trucking to bankrupt most of America's freight railroads in a matter of decades. But the self-driving pioneer's point was, I suppose, that in the future when you want to change where this "train" goes you don't have to lay new track—you just slide the "rails" around on a screen.

Levandowski didn't realize how prescient his metaphor was. Lots of other companies were also working on self-driving trucks. Otto certainly wasn't the first. But one reason why other companies lagged behind the American startup was that they were focused on the more difficult task of actually linking computer-driven trucks into virtual "trains," a close-following formation that is known in the business as a *platoon*.

Earlier in 2016, a half-dozen truck convoys had converged on Europe's busiest port, in the Dutch city of Rotterdam. But this was no marketing ploy or DARPA road rally. Rather, it was a full-bore display of the EU's emerging industrial policy, which hoped to tap European manufacturing giants to speed the flow of freight while also reducing greenhouse gas emissions from trucking. The virtual version of a conventional tandem tractor-trailer deployed by Scania and Daimler—respectively, the world's largest truckmaker and its second-largest manufacturer of buses—performed flawlessly, as the trucks' self-driving software coupled up via a high-speed wireless network instead of a hitch. Drafting in each other's slipstreams, the platooned vehicles were more than 10 percent more fuel efficient than each rolling down the road alone. By 2019, Scania success-fully tested platoons with as many as four trucks moving together as a single unit, and clever enough to adjust spacing if a passenger car slipped into the 35-foot gap between them.

There's a long history to the kooky idea of tying trucks together into a single motorized column—during the Cold War, both the Americans and the Soviets used "land trains" to supply remote Arctic bases. And in Australia's Outback today, "road trains" are still commonplace, carry-ing fuel and bulk materials over long distances. But despite platooning's potential, its business case isn't nailed shut. In 2019, Daimler suddenly got cold feet, claiming that the efficiency gains offered by platooning weren't enough to offset the extra cost of self-driving equipment. We may need to wait awhile longer before we see these physical couplings severed by software—in trucking at least.

But as often happens with new technologies, platooning may yet become the best solution to an altogether different problem. Over the last two decades, bus rapid transit (BRT, for short) has helped cash-strapped cities around the world roll out train-like transit capacity for a fraction of the cost of building railroads and subways. Following early success in Brazil in the 1970s and 1980s, the idea has spread widely—today more than 33 million people in 195 cities worldwide use it daily. Further gains for BRT may be elusive, however, as the scheme has reached an unyield-ing capacity plateau. The most highly geared systems, like Bogotá's TransMilenio—which, at rush hour, can have buses running as close as 60 seconds apart—can move more than 40,000 passengers per hour. Yet that's half or less than the throughput of the world's busiest subways in cities like New York, London, or Hong Kong.

One upper limit on BRT capacity is the feasible size of human-driven buses. The largest buses in use today are Curitiba's serpentine biarticu-lated models. These giants, which ply express routes through the Brazil-ian city that sparked the BRT craze more than 50 years ago, use flexible joints to link three coaches and can carry more than 250 passengers. But if self-driving technology can safely link up trucks into virtual convoys, could it also bond buses together for high-speed transit—literal *software trains* (Figure 3-3)? The orderly, efficient flow of platooning passenger cars was, after all, one of autonomists' greatest dreams for the driverless revolution. So why not supersize that computer-choreographed game of follow-the-leader?

If proven practical, software trains would smash the limit on BRT capacity, giving the approach a much-needed boost. It's easy to imagine self-driving bus convoys as spacious as a subway—six, seven, or eight buses in length. While you'd lose some capacity by doing away with the open gangways that allow passengers to move between coaches in today's articulated buses, you might still double the number of people carried by today's biggest behemoths.

What's more, automation could significantly improve the comfort and convenience of big buses. Software trains wouldn't be as restricted to special roads as conventional articulated buses—they could momentarily separate to bend around tight corners on city streets. Meanwhile, with an autopilot providing precision lateral control, dedicated busways could be narrower. That would make it easier for cities to splice them into existing road corridors. "Dual-mode" operation, where coaches originate at different points but link up for more efficient travel on a busier trunk line—impossible for conventional BRT—would also be a breeze. After passing

Figure 3-3. *Rovers and software trains.* (a) AV wheelchairs provide indoor and outdoor mobility for nonambulatory people. (b) Cargo bikes become electrified, and automated, providing low-emission, safe, and quiet mobility for human-assisted deliveries. (c) Scooters employ limited automation to deadhead between rides and return to charging points overnight. (d) AV hoverboards store themselves in back alleys and under parked cars, and can be hailed electronically on a moment's notice. (e) AV tricycles provide high stability and various operating modes: fully powered or pedal-assist, self-driving or manually steered. (f) Software train in use today, developed for high-efficiency platooning of AV long-haul tractor-trailers. (g) Future application of software train to mass transit, by platooning of AV buses. DASH MARSHALL.

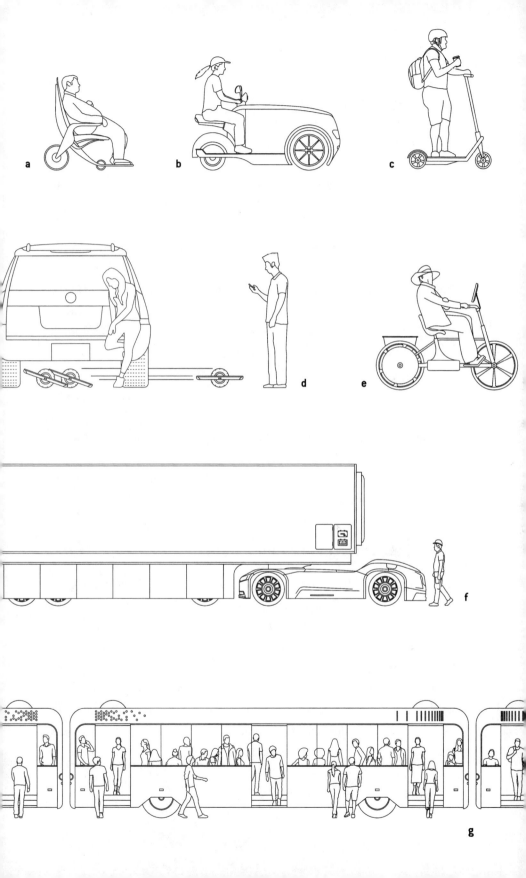

a

b

c

d

e

f

g

through the city center, the buses might split up again to increase the number of destinations served with a single-seat ride. Finally, with cool-headed software taking the place of brake-pumping drivers, rides could be smoother and more like the comfort of riding a train.

The first step toward making software trains real is automating individual, full-size city buses. Mercedes-Benz, where Daimler's busmaking group sits, has committed more than $200 million to an AV system it calls CityPilot, targeting 185 BRT systems worldwide to upgrade more than 40,000 buses already in use. But there is growing interest in linking the self-driving buses of the future into real-life road trains. In Austin, Texas, the scheme has already shown up in a forward-thinking regional transportation plan. "It's not some sci-fi vision," the Capital Metropolitan Transportation Authority CEO, Randy Clarke, told the *Austin Business Journal.* "It's much closer than many people realize."

Packing people by the hundreds into high-speed vehicles driven by computer will require a new level of confidence in driverless software. But by the millions we happily board planes flown by autopilot every single day. If we're to meet the challenge of moving huge numbers of people while cutting the carbon emissions generated in doing so, software trains are too tantalizing an opportunity to ignore.

Civic Caravans and Urban Ushers

I'm a huge fan of IKEA. The company's soft sell of egalitarianism, good design, and healthy living is such a refreshing alternative to American consumerism that I sometimes stop in just to have lunch. My local blue box is smack up against a massive container port, so it's a bit like taking a holiday in a Scandinavian-run free-trade zone. As I stroll through the tasteful, affordable kitchens and living rooms—many of them hewn from sustainable lumber and bamboo—I think about how much I would love to live in an IKEA-designed town. And so, when the geniuses at Space10, the company's Copenhagen-based design lab, came up with a scheme for future living that blurred the boundaries between vehicles and buildings, I took notice.

"We don't have ambitions of manufacturing cars," IKEA's Göran Nilsson told *Quartz*, an online magazine. "But in a future where people no longer have to worry about driving, vehicle interiors can expand to a point where we no longer are designing cars, but rather small spaces. Then it's suddenly an area where we have a lot of experience." As with so many of the company's best innovations, it was a single, simple idea that gave birth to a whole collection of new products. Space10's designers started with a basic AV chassis about the size and shape of a food truck, fitted out with the same sort of pop-up and fold-out appendages those vehicles employ to provide shade, signage, and counters. But then they began to iterate on the form, fitting these driverless structures out as government and community facilities where people might go for their everyday needs. In one, they put a health clinic to reach those who can't get to a doctor; in another, a farm stand to bring healthy produce to inner-city food deserts.

These clever mash-ups raised a much bigger question. Why go to a government building when the building can come to you? The IKEA crew members were onto something. But they got a bit carried away with the technology. Like autonomists are wont to do, they wrote human beings out of the picture completely. They replaced the government workers inside their self-driving trailers with screens.

———

THERE'S A BETTER WAY to exploit building-sized AVs. What if *civic caravans* (Figure 3-4), as I call these roving offices, could bring teams of public servants closer to the people most in need? Instead of removing people from the equation, automation would empower them to do their jobs more effectively.

Imagine the public square of the future. A new day dawns, and a herd of these behemoth vehicles have arrived overnight. As their parking sequence winds down, the full might of their self-driving supercomputers are retasked to the business of helping residents. Screens, inside and out, spring to life with information, services, and public art. Unlike government offices today, full of complicated forms and long queues, the front lines here are staffed by avatars, which—like C-3PO from *Star Wars*—are able to communicate in any and every human means known. They're clever enough to handle most routine requests, which frees up human staff, who are now here in your community instead of miles away, allowing them to focus on the challenging and difficult cases that need close human attention and judgment.

Mobile buildings are an ancient idea, dating to at least Babylonian times, when siege towers were first rolled up to breach fortress walls. And the idea of roving, robotic edifices is a product of the 1960s, when British architect Ron Herron sketched "Walking City," a vision of insect-like robots dozens of stories tall roaming about the earth in search of resources and trading partners. The proposal has been called "the international icon of radical architecture of the Sixties." Today, the closest experience you can have of what a building-sized vehicle feels like is at Concourse D at Washington Dulles International Airport, where 36 massive "mobile lounges," designed a half century ago by Eero Saarinen, still ferry passengers from gate to plane. Each weighs some 76 tons and carries up to 90 people.

In the future, self-driving vehicles could grow to gargantuan proportions. They could move at night with little cost and minimal disruption to human activities, perhaps even communicating directly with smart traffic signals to open and close roads as they go. When the space shuttle *Endeavor* was moved across Los Angeles in 2012, the 12-mile journey required a custom-built 160-wheel carrier and hundreds of human escorts, at a cost of more than $10 million. But could such a heavy lift become an inexpensive, routine, automated operation in cities of the future?

Civic caravans wouldn't simply be self-driving versions of prefab government trailers. Computer vision, at such low speeds, wouldn't be used just to look for obstructions *ahead*. Its gaze could be turned *down* as well, collecting precise imagery of potholes and road conditions. Pairing such superhuman sensing with active, computer-controlled suspensions could create a mobile platform as stable as the ground below—allowing delicate, light, and airy structures of metal and glass to rise above. Civic caravans could well be the defining form of public architecture in the early twenty-first century.

Civic caravans could also reshape local governments from the inside out, helping prepare them for the stresses of the twenty-first century. Cities today struggle to deliver as many as 1,500 services to citizens every day, according to Citymart, a group that helps cities innovate. These range from maintaining infrastructure and building schools to monitoring public health and administering justice. Today, most local governments separate these functions into different departments that don't always work together. People in need often fall between gaps. Big opportunities for efficiency are missed. But as an itinerant alternative to City Hall, civic caravans could house tactical teams that streamline and integrate service delivery by combining specialists from many different agencies.

Working this way would be transformative for government and citizen alike. You wouldn't have to stand in line, or even go online, to get the services you need. Instead of dealing with dysfunctional bureaucracies, you could take your child to a civic caravan close to home and deal with all the educational, health-care, and recreational needs in one place, where the staff knows your face. Gone would be the central facilities—and their costly upkeep.

Gone too would be the predictable consistency of civil-service work. Civic caravans' staff would have a measurable stake in how things turned out for the people in their care, and they'd be held accountable to it. But they'd have a greater, more visible impact. Government work would be more effective, rewarding, and meaningful than it is today.

Whether the caravans stayed a day or a week would depend on how quickly they achieved results. But they'd never show up randomly. They'd always be on the lookout for weak signals of neighborhood socioeconomic

a

b

c

d

distress in the city's stockpile of data. They'd constantly be weighing the impact they were making here against decisions about where to decamp to next.

———

WE'VE COME A LONG WAY and there's one more species of AV to see. This breed of AV is relatively small. You'll find them swimming remora-like alongside those massive caravans, and they could prove just as destabilizing to the municipal order. Like caravans, they blur the lines between familiar categories. I call them *urban ushers* (Figure 3-4). Part vehicle and part robot, they'll be the future's beat cop, parking meter, stop sign, and traffic signal all rolled up into one.

The easiest way to understand the opportunity for ushers is to stroll down the nearest sidewalk and take a mental inventory of the "street furniture." That's the urban planners' term for all the crap that's cemented into the ground—the mailboxes, traffic signs, lampposts, and so on. Now imagine all that on wheels.

If it sounds crazy, consider how much public benefit self-driving street fixtures could deliver. For starters, urban ushers would help us clear sidewalks of obstacles. Today, we navigate the accretion of a century's worth of fixed applications when we perambulate. A single usher might replace a dozen pieces of equipment that clutter the sidewalk.

But ushers could also sweep away equally rigid and often cumber-

Figure 3-4. *Civic caravans and urban ushers.* (a) Caravan containing a community health clinic moving to a new neighborhood at night, in a coordinated redeployment on dynamically closed streets. (b) Patrol usher responds to a health emergency by summoning assistance, securing a cordon around people in distress, and providing basic triage and medical instructions. (c) Street-steward usher monitors and controls traffic through wireless coordination with AVs, and sensors provide ongoing monitoring of air quality and noise. (d) Street-steward usher dynamically reallocates streets, directing traffic and pedestrians with wayfinding signage, real-time updates to navigation apps, and speech in multiple languages. DASH MARSHALL.

some rules that govern the public right-of-way. They could direct other AVs wirelessly and use voice and gesture to communicate with pedestrians, cyclists, and human drivers. "Block captain" ushers could be tasked with neighborhood ground-traffic control, tracking every vehicle that comes and goes, making sure that children all get safely escorted across the street, and managing curbside parking for deliveries, taxis, and passenger shuttles.

Ushers would take on temporary, specialized tasks too. They'd activate onboard screens and become helpful information kiosks for residents and visitors alike. They'd close streets for block parties and organize extra seating for strolling seniors on Sundays. In an emergency they could swarm in groups to light evacuation routes or form a blockade.

Over time, as ushers brought about a new level of precision control over the public realm, we'd start using them to make more radical and permanent changes. After allowing their AI to observe us in excruciating detail, we might unleash it to make preemptive changes of its own, like sequestering streets for exclusive pedestrian use or reducing the flow of traffic. And everywhere, I suspect, we'd look for opportunities to monetize the spaces they managed. But we'd have to be careful about unleashing supercomputer-powered meter maids on an unsuspecting public.

Ubiquitous urban ushers would bring about a degree of street-level sensing that would shock us all. But would that mean the creation of a new kind of sidewalk surveillance state? Possibly. The answer would depend on who controlled them. It's easy to imagine all kinds of uses where ushers could collect vital data to protect the public interest. For instance, childhood asthma disproportionately harms disadvantaged groups. But most cities have relatively little data on ground-level ozone pollution, a major cause of this condition. Fitted with the right kinds of air-quality sensors, ushers could give cities a much clearer picture of the problem. But if ushers were not operated by government, ownership and reuse of their multiple data streams would be difficult to control and controversial.

Either way, ushers would be less Big Brother–like than the current default approach to "smart cities"—blanketing entire territories with a panopticon web of pole-mounted sensors. In Japan, Hitachi has demonstrated an experimental penguin-like "patrol robot." This adorable droid trundles along footpaths, greets schoolkids and seniors by name, provides auxiliary street lighting at night, and flaps its flipper-like arms while summoning paramedics wirelessly when it detects a medical emergency. Snoops like this can't hide. You'll know when they're watching you.

A Robotic Role Reversal

"The robotization of the city" is just beginning, muses Paris deputy mayor Jean-Louis Missika, whose purview includes architecture, urbanism, and economic development. Indeed, even our explorations here have barely begun to imagine the full variety of the menagerie of automated vehicles to come.

What about motorcycles, for instance? Automation will be far more transformative for these cheap and fast but dangerous devices. And if you delegate the balancing act of a two-wheeled speed demon to software and gyroscopes, many more people might downsize to high-revving, lane-weaving hot rods. Highways could carry far more single-passenger vehicles.

Then there are the hominids, which are blurring the line between driverless vehicle and autonomous machine. Boston Dynamics' Big Dog was built as a cargo-carrying support vehicle for troops. It can carry a 100-pound load over the roughest terrain. MIT's Cheetah 3 is superior to many AVs, since it can operate in complete darkness or torrential rain. Instead of cameras, radar, or lidar, it navigates entirely through tactile feedback picked up by its four feet—a technique its creators call "blind locomotion." For now, both bots are strictly limited for military and industrial use. But is there any doubt these technologies will soon find their way into our world?

Don't get me started on self-driving tractors.

Where is all this specialization taking us? In his history of the car, *Autophobia*, Brian Ladd notes that "a striking fact about the automobile ... is how little it has really changed in the course of a century and more.... A wide variety of motorized and wheeled vehicles have clattered and dashed across the surface of the earth, but the great majority of them are recognizably of a single type: a metal box, mounted on four wheels, powered by a petroleum-burning engine." Automation is going to break that box wide open.

Don't take this as a prediction that cars will disappear. They'll be reinvented, too, to become more useful than ever by fueling and parking on their own and taking themselves to the car wash. But compared to today, they'll have a lot more competition for our transportation dollar and play a greatly diminished role in how we get around. Because wherever you want to go and however you want to do it, a specialized AV alternative that costs less, performs better, and spews less carbon will be but a beck and call away. Or, perhaps, it may already have arrived, having anticipated your need.

Beyond this, all bets are off. Designers will dismantle the barriers separating vehicles, robots, and buildings. Fixed will become mobile. Insides will become outsides. Passive will become interactive. And by 2050, we'll have invented far better, heavily automated ways to move billions of people with none of the danger and a fraction of the cost and fuel we use today. In the process we'll discard some of our most deeply held assumptions about the structure of government, the purpose of architecture, and how much of our own human locomotion is done by muscle—or by machine.

But to get there (and I believe it's a place we want to go), we have to overcome a huge mental roadblock—our outdated cultural understanding of automation itself. In school, computers are explained as reducing-machines that distill everything they encounter to a sequence of ones and zeros. In movies, intelligent machines suck the richness out of harmonic speech and replace it with droning synthesized voices. We even have a dance that mocks machines' feeble, jerking attempts to emulate our graceful, fluid movements—the Robot. Everywhere we turn, we're told that automation is fundamentally about *getting rid of* variations.

That stereotype of automation is changing fast, and in the future I believe we'll come to see intelligent machines as a technology that *accentuates* differences rather than eliminating them. Today, AI is making our online experiences more responsive and personal. AVs will carry those capabilities out into the physical world with every trip. And in doing so, they'll serve us in better and more varied ways than the vehicles of the past.

This robotic role reversal won't be easy. We'll need to learn to trust vehicle automation enough to explore the possibilities. No doubt, we'll face trade-offs along the way. But this leap of faith could bring about an extraordinary turn in our favor. Exploring the full range of inventions this technology revolution offers may allow us to finally cast aside the automobile and its insatiable appetite for fuel, land, and human sacrifice.

And that would make all the difference.

4 Reprogramming Mobility

There are still millions of people who want to drive but lack the elemental courage. For example, the blind, the aged, timid women, cripples, little children too young to secure permits to drive. There is a market for you to think of.

—"The Living Machine," 1935

In the 1930s, the future felt close, yet still out of reach. Space-faring ships, mobile phones, and laser guns came to life on the pages of science fiction monthlies. As our great-grandparents imagined much of the technology we now take for granted, these tales provided relief from the despair dished up as the Great Depression dragged on.

Self-driving cars saw their share of ink, too. And one story published in 1935, "The Living Machine," predicted many details of autonomous vehicles that wouldn't be realized for another 85 years. Gracing the pages of *Wonder Stories*, this was a rags-to-riches tale. A young inventor, John Poorson, has been struck by a reckless motorist. The glancing blow "injured little except his pride and clothing," but he "had a single idea; it had been knocked into him." A year later, the genius emerges triumphant from self-imposed seclusion in his workshop with a mysterious, shining, eight-inch-diameter sphere. Much like a young entrepreneur today, he's eager to sell to the highest bidder. The pitch is tense:

> The little inventor was the only calm person in the room. At last Babson recovered himself enough to ask:
>
> "You mean that you have given [the car] a brain?"
>
> "Something like a brain. You know it had a nervous system before, electric wires and all that sort of thing. What I have done is furnish a sort of clearing house to receive impulses, place a value on them and sufficient reactions so the car can take care of itself."

The prescience of this passage is uncanny. Poorson's "brain"—a "clearing house" built to "receive impulses" and "place a value on them"—is as good a literary take on the deep learning networks behind the wheel of today's AV as any. What's more, Poorson's device is housed under the car's seat—the same spot where today's self-driving supercomputers live. And much like modern AVs' brains, the inner workings of his supernatural sphere are a closely guarded trade secret. Even as readers, we never learn what makes it tick.

More intriguing, however, is how "The Living Machine" describes the social and psychological upheaval accompanying driverless vehicles. The author, David H. Keller, was a practicing psychiatrist, a profession that gave him a talent for extrapolating predictions of future human behavior. "Old people began to cross the continent in their own cars," he writes. "The blind for the first time were safe. Parents found they could more safely send their children to school in the new car than in the old cars with a chauffeur." Nearly a century before Google's 2014 launch of a self-driving-car prototype—a public-relations stunt featuring seniors, disabled people, and kids being ferried about on their first driverless rides—Keller foretold almost to the letter how it would occur.

As the story continues, Poorson's invention sweeps the nation with a speed that would impress investors today. Nearly all cars are converted to automatic control in five short years. But events take a dark turn. Even as this self-driving paradise takes shape, it becomes a killing field. In a gruesome twist, a cocaine-tainted batch of gasoline turns the robocars against their human masters. "There seemed to be a personal devil in each of them, directing their actions," Keller writes. "Cars, without control, coursed the public highways, chasing pedestrians, killing little children, smashing fences.... Fifty million machines were on a wild riot of uncontrolled destruction." The renegade robocars even discover how to gas themselves up at service stations, perpetuating the reign of terror indefinitely.

For all its B-movie sensationalism, the lessons of "The Living Machine" remain fresh and relevant today. AVs will deliver mobility to millions left behind by conventional cars. But the driverless revolution doesn't begin and end with vehicles. "Traffic was paralyzed," Keller writes. "The nation became panic-stricken. Schools were closed." The impacts will reach much further—full automation will transform the entire transportation *system*.

Gridlock in Tel Aviv

Transportation-system breakdowns are now so common and colossal they have an internet meme all their own: the *trafficgeddon*. To make one, simply take the name of the road, tunnel, or train that's out of commission and splice it with *Armageddon*. There was Los Angeles's Carmageddon, the 2011 closure of the 405 freeway; and Atlanta's Snowmageddon, a 2014 blizzard. (Seattle's Viadoom, the 2019 permanent closure of Seattle's Alaskan Way Viaduct, didn't lend itself linguistically to the pattern, but was still a rather effective exception to this new rule.) The insinuation isn't subtle—what you're saying when you construct such a Frankenterm is that the event is a catastrophe on par with the end of the world, the final battle between good and evil foretold in the New Testament's Book of Revelation.

Perhaps it's fitting, then, that one of the most stunning trafficgeddons to date took place in Tel Aviv. After all, the Israeli capital is little more than 50 miles from Megiddo—the ancient, ruined city that true believers expect will be the actual site of Armageddon.

That summer weekday in 2011 began as usual. Millions of Israelis awoke to greet a new day. In car-dependent Tel Aviv, a difficult morning lay ahead. On average, more than 60 percent of commuters in the Israeli capital region drive to work each day. But many drivers had a new tool to navigate the gridlock. They were connected to a new social network called Waze. While other apps collated sensor data on traffic conditions and pushed reports back down to commuters, Waze prodded you to play cub reporter, too—tapping out the specific cause of a slowdown (a crash, debris in the road, etc.), the location of police speed traps, or other tidbits of information of interest to fellow travelers. The whole thing was riddled with cute, anthropomorphic icons and emojis. You joined tribes and earned points. Driving with Waze felt more like a game than managing a commute, and it could be addictive. These crowdsourced congestion reports quickly became an indispensable tool for hundreds of thousands of Israelis hacking their way through the city's heavily congested road network.

But that morning, something was wrong. As rush hour got underway, Waze's maps were blank. Meanwhile, seven time zones to the west, in the Washington suburb of Ashburn, Virginia, all hell was breaking loose. Like many startups, as Waze grew, the company had rented virtual servers from Amazon's Web Services division (AWS). Today, AWS offers a planetwide grid of reliable and redundant server farms, and that infrastructure powers vast swaths of the web. But nearly 10 years ago, the cloud was far less robust. A troubling series of outages had plagued Amazon that year, but the one on August 7, 2011, was particularly disruptive. That night, a line of thunderstorms rolled through Ashburn late in the evening, down came a tree, the tree fell on a wire, and out went the lights on AWS . . . and Waze.

Back in Tel Aviv, rush hour was a mess. Israeli friends tell me the gridlock that day was the worst they'd ever seen. Drivers who'd come to rely on Waze for both directions and traffic alerts had to fall back on their own wits. And when roads are as saturated as Tel Aviv's, carrying far more traffic than they were designed for, all it takes is a tiny disruption to congeal the smooth flow of vehicles into a sticky mess. Waze followers, in their sudden bewilderment, brought traffic to a crawl for the majority.

At first, I judged Waze harshly. What hubris these technologists had, tinkering with the daily decisions of millions of people, upsetting the dynamics of systems they couldn't possibly understand. But the more I thought about it, the more I realized this was the point. In a short period of time, Waze had wired itself into the Israeli transportation system so thoroughly it could no longer be removed without catastrophic effects. Until that day, no one had noticed—not even the Jewish state's vigilant defense planners. But after August 2011, Google sure did. Not long afterward, the search giant acquired Waze for more than $1 billion, the biggest such deal in Israeli startup history.

Waze's infiltration of the Israeli street grid wasn't an isolated incident; it was a glimpse of a much larger process of *reprogramming mobility* that's playing out all over the world. And in the months that followed, as I continued to travel around the world, I saw the same pattern repeated everywhere—infrastructure, vehicles, and people's personal devices... ever more connected and coordinated by software. In New York, I got nailed with a 500 percent surge charge from Uber that left me speechless. Yet I found it remarkable that a company's code imposed its own form of demand-based transportation pricing in a city that had itself tried but failed to do so for decades. In Argentina's second city, Córdoba, I peered over the shoulders of traffic engineers routing buses around a street protest via the city's fully programmable traffic-signal system, Latin America's first. And in San Francisco, I downloaded Serendipitor, an app that let me reprogram *myself* by charting a random walk through the city's grid.

All of these innovations were variations on the same theme—a common understanding that we already have plenty of transportation and just need to use it more effectively. Instead of pouring concrete, we could expand transportation systems by investing in the digital realm. As if on cue, in 2017, the city of Los Angeles published a sweeping vision for future transportation, *Urban Mobility in a Digital Age*, giving this strategy the power of law in the world's most influential center of car culture.

The reprogramming of mobility doesn't require automated vehicles—the effort is already widespread and well underway. On the contrary, reprogramming mobility is setting the stage for driverless technology's rollout. What this trend tells us is that we've been way too focused on the hardware of AVs and future streets, and the software *inside* these machines—and ignored the software at work *outside* them. Tel Aviv's gridlock highlights how the code that does the dispatching, which is not in the vehicle, will shape our world more profoundly than the software that does the driving. It's what will tell AVs where to go, when, and how much to charge.

This was a revelation for me. And like Neo at the end of *The Matrix*, once the walls were stripped away (well, in this case, the pavement), and the digital ghosts that shape our every move were revealed, it was impossible to unsee them.

At Your Service

Electric streetcars swept into American cities with astonishing speed during the closing years of the nineteenth century. The first practical demonstration of the new technology took place in 1887 in Richmond, Virginia, and within ten years "electric traction" had completely supplanted horse-drawn railways across the US, Canada, and much of Europe too. Much like Uber and Lyft today, streetcar companies were backed by deep-pocketed investors. And like their latter-day counterparts, they, too, quickly grew too big to fail.

Electrification unleashed enormous pent-up demand for urban mobility. In Philadelphia, 75 percent of the streetcar network was electrified by 1895. The number of streetcar passengers citywide doubled between 1885 and 1895, to more than 220 million trips a year. But the expansion was hasty and uncoordinated. The streetcar boom became a crisis. As the web of routes grew, competing lines crisscrossed with little rhyme or reason. On some streets as many as three sets of tracks and overhead power lines ran in parallel. What's worse, every leg—even those involving a change between streetcars owned by the same company—required a separate ticket. In the early 1890s, the majority of Philadelphia's streetcar commuters paid at least two fares to get to work.

It would fall to Peter Widener—a butcher by trade, a political insider by cunning, and one of the city's aspiring transit tycoons—to bring order to this tangled mess. By 1895, his Philadelphia Traction Company (PTC) threatened to dominate the city's transportation business. During the previous decade, PTC's ridership had tripled as it gobbled up rivals. But that year, Widener introduced a change in ticketing that would have profound consequences. Suddenly, PTC streetcars would take Philadelphians anywhere they wished to go for the price of a single fare.

The *free transfer*, as it was known, was a simple service innovation in an age of great engineering achievement—and was, to my knowledge, the first offer of its kind anywhere. But it changed everything for Philadelphia. The riding public flocked to PTC's streetcars, which for the first time functioned as a cohesive transit system in the modern sense. The move proved so popular that not only did it finance itself, bringing in more in new customers than it gave up in forgone fares, it gutted the competition. More than half of those using transfers were new customers poached from rival streetcar lines. In short order, PTC took control of the remaining independent streetcar lines, killing them off or integrating them into the citywide network. Within a decade, the company's monopoly was secure.

———

TICKETS REMAIN AN UNLIKELY PLACE to begin building a better transportation system. But today, smartphones are displacing paper stubs— Philadelphia's mass-transit agency, SEPTA, made that leap in 2018—and the way we buy mobility is changing faster than it has in decades.

Consider the choices you face today when traveling from one side of a city to another. After deciding on your destination, you move on to picking a "mode," as transportation geeks call it. There could be a half dozen or more to choose from, including private cars, private bikes, taxis, buses, trains, trams, shared bikes, shared cars, and simply walking. Layer in all the different pricing options—pay-as-you-go, daily passes, monthly passes, and so on—and deciding how to get to work suddenly requires a master's degree in economics.

Consider my morning commute crossing the Hudson River into Manhattan, for instance. On any given day I can choose between a subway, a bus, or a ferry. Deciding which to take is a complex calculus of cost, comfort, and convenience that changes depending on work and family schedules, the weather, and service disruptions. Thankfully, there are loads of apps to help tame this tangle of consumer choice—it's one of the busiest bits of reprogramming mobility going on today. Google Maps, CityMapper, Transit, and a number of similar apps happily identify the fastest, shortest, or cheapest route. Some even provide tools to sort trip options on carbon emissions or calories burned.

With luck, I've now decided *how* to go. Now the most difficult part begins. It's time to pay. Today, smartphone in hand, one can walk up to most any bus, train, or ferry anywhere on the planet, buy a mobile ticket, and walk on. In a growing number of cities, that one e-ticket will work as you move across railway, metro, and bus, kind of the way the Philadelphia free transfer did so long ago. But in many other cities, like New York, you still have to pay most transit fares separately. My experience is extreme, but it illustrates how bad this ticket tango can get. I sometimes buy three different tickets to get to my office in Manhattan in the morning—one each for a streetcar, a subway, and a bus. I spend half the trip juggling kids' backpacks and my phone, frantically checking weather, arrival times, and trying to purchase tickets on buggy, badly designed transit-system apps.

Wouldn't it be better if I could plan my trip and buy my tickets in just one app? Even better, wouldn't it be best if I could do it in the app of my choice? That's the big idea behind a new approach to ticketing known in the business by the mouthful *mobility as a service*. If you can't remember all that, just use the nickname that insiders do—*MaaS*. Pronounce it like a Boston brahmin describing the sphagnum in their garden ("mahss"), and you'll nail it.

The problem MaaS tries to solve is the same one that Philadelphians faced all those years ago. How do you glue a bunch of disparate transportation lines together into a cohesive network? You can think of MaaS as a modern-day twist on Widener's free transfer—but rather than consolidating ownership of the entire system, MaaS seeks to spur competition by building an internet-connected bazaar instead. Anyone with services to sell can advertise, and anyone can buy. Software does the work of stitching all those à la carte tickets together into a single, tailor-made digital pass that's good for your entire trip.

To see how it's supposed to work in practice, let's time-travel a few years into the future and fire up my favorite MaaS-enabled app, Transit, which is made by some clever folks in Montreal. Say I'm in town and I want to travel from Trudeau International Airport to McGill University. I punch my destination in and Transit pulls in available services advertised on the MaaS bazaar that match my time and route. I can do an apples-to-apples comparison and see whether Uber or Lyft offers a better ride-hail deal on this run right now. But I can also compare apples to oranges and see how the price and travel time for a city bus compares. (In Montreal, you'll quickly discover that the awesomely named 747 downtown express bus will save you a bundle.)

Behind the scenes, all hell is breaking loose in the cloud. Mobility providers jockey to underbid each other inside the MaaS bazaar, for the privilege of being displayed on my screen's top shelf. Third-party brokers try to elbow their way in with discounted ticket bundles they've assembled to more precisely target my niche. Finally, the results are weighted according to my personal set of filters that have nothing to do with money—maybe I like green modes of travel or minority-owned businesses and want to boost them in the results. My selection made with a tap and a swipe, the trips are booked, my account is charged, and—much like in an app store today—the MaaS middleware does the dull accounting to make sure everyone gets paid their fair share for hauling me across the city.

Easy, right? I'd take this any day over my current morning routine. And that's the whole point. But why the ungainly term, *mobility-as-a-service*, that so deftly obscures its potential? It's an ugly bit of industry jargon to be sure, but it does highlight two essential innovations. Let's unpack the phrase—it will help reveal these bigger ambitions.

First, why call it *mobility* and not *transportation*? Transportation is tangible. It's something you can put your hands on. It's roads and rails and buses and trains. Mobility is sort of soft and uncertain. Say *transportation* out loud, and you're talking about "a system or means of transporting people or goods." That's the supply side of the market—the realm of capitalists, unions, and government agencies that move us where we need to go. But utter the word *mobility* instead and you mean "the ability to move or be moved freely and easily." Now you're talking about us, the traveling public. This is a massive shift in attitude, and it seeks to shape future innovation by putting customers' needs foremost. And it opens our minds to other means of achieving that free and easy movement. It doesn't all have to happen by cars, buses, and trains. When you change the subject from transportation to mobility, you can instead simply make it easier to walk, bike, or even "travel" to meetings by videoconference instead. *Transportation* assumes that vehicles are the answer. But we can create lots of *mobility* without building anything new or burning any more fuel.

The second part of this grand strategy, *as-a-service*—what's that gobbledygook? It sounds odd, but it's surprisingly straightforward. This is a recipe for reorganizing the remaining people-moving we can't avoid after we make the shift from transportation to mobility. It's an approach that we're all familiar with, too, as so much of our online lives is already delivered as-a-service. Take webmail, a good example of software-as-a-service. Twenty years ago, if you wanted to read email you installed an app on your computer and it fetched the messages from a server. As your little digital letters were plucked from the remote mailbox, they'd disappear over there and appear here. It was an utterly useless anachronism, which webmail thankfully did away with by moving both the email software and the data back to the server for good. You no longer had to worry about which computer you had opened your emails on earlier. Now they were always accessible, from anywhere in the world. It didn't take long for many of us to realize that if you can get your email through any web browser, why own a computer at all? The age of the PC was over and the era of the cloud had begun. As a result, the internet now brings a wider variety of more useful services to more people at less cost than ever imagined.

I find it easier to understand MaaS by dropping the hyphens and rephrasing it thus—*mobility, as a service*. Let the comma linger, and the meaning will sink in. The ability to move, delivered to you on demand. This is perhaps the greatest promise of reprogramming mobility—and as we began to see in the vehicular variety of Chapter 3, automation will bring it forward with speed.

Making the most of MaaS, however, will force us to confront issues that have dogged digitalization within other sectors. As we expand access, how do we ensure it is done fairly? As we drive innovation, how do we decide when more consumption is too much? How do we deal with the valuable, and dangerous, torrents of data that these technologies will unleash? And lest we forget the lessons of the past, while these ticketing innovations may pave the way toward a more seamless traveling experience—will they once again catalyze consolidation instead of competition?

Let's explore two possible futures, showing how mobility as a service might exploit self-driving taxis and driverless shuttles. These examples illustrate what's possible when MaaS maximizes competition and choice. But they also highlight why strong policy is needed to guide markets that deliver not only scale but sustainability and inclusion as well.

Taxibot Takeover

Among its well-aimed takes on the future, "The Living Machine" predicted precisely how automation would transform the big-city cab. "As a driverless taxi was finally introduced," Keller wrote, "it was a relief to enter one of the comfortable autos, whisper a direction down the tube, and just know that you would be taken to your destination in the quickest, safest, and cheapest manner." We're just a few years away from this becoming a widespread reality. But we'll utter our wishes into a phone or a concealed microphone instead. More surprising, we'll get a response, too. The same computer that does the driving will synthesize a soothing, even seductive, voice to clarify and confirm our wishes.

AV cabs, or *taxibots* (the preferred term of the Organization for Economic Co-operation and Development, or OECD), will wow us with convenience (Figure 3-2, page 61). Calling cabs with such ease is sure to catch on quickly. But the real draw for automated taxis will be their rock-bottom price. Today, drivers take home about 80 percent of ride-for-hire fees. But taxibots will all but eliminate the share that goes to labor. They will also reduce costs in other ways. Today's cabs spend half of their working hours empty, not earning a penny. By eliminating bathroom breaks and shift changes, and by reducing both cruising (looking for passengers) and deadheading (driving to respond to an electronic hail), taxibots could cut operating costs by a full 75 percent over human-driven vehicles—from about $2.50 per mile to under 70 cents.

As economies of automation kick in, the impact on the taxi business will be explosive. In 2018, ride-hail operators grossed $5 billion worldwide. Powered by cheap automation, the take could grow to $285 billion annually by 2030, according to Goldman Sachs. Two trends would drive the expansion. Lower fares would make taxibots cheaper than the total cost of owning and using private cars on a per-mile basis, and larger taxibot fleets would steal riders from public transit by providing door-to-door rides at a competitive price. By 2040, it would be hard to find a human-driven taxi in a big city of the global North. By 2050, as the market grew into the trillions, more than *one billion* taxibots could ply the world's streets.

Now, this tale of a taxibot takeover makes for compelling copy. It's a favorite of autonomists' fireside chats. But it makes the same fundamental forecasting error we've seen time and again already—it underestimates the sheer variety of the future. Why do we presuppose that taxis will remain a singular thing, the same sort of common carriage that the beloved yellow cabs of New York or the black cabs of London provide today?

No assumption about the future could be more mistaken. Like everything else in the driverless future, taxibots will defy our expectations. They'll strategize, specialize, and discriminate to serve the customers they want to capture. The number that will define taxibots' future isn't how *many* cabs there will be, but how *many kinds*. Because automation—even as it expands the market for taxis—will change what they do.

THE HYPERSPECIALIZATION OF RIDES is already well underway. Beyond the ever-expanding array of choices inside the ride-hail apps like Uber and Lyft—big cars, green-energy cars, wheelchair-accessible ones, and so on—independent carriers are finding countless niches within which to carve out viable lines of business. HopSkipDrive provides supervised mobility for schoolchildren. SilverRide targets senior citizens, some 600,000 of whom give up driving every year in the US alone. And New York City's Pet Chauffeur does $1 million in business annually and has "transported everything from leopards to bulk shipments of lab rats."

This long tail of ride-hail doesn't compete on cost but on service. HopSkipDrive emphasizes the security screening, credentials, and personal warmth of its "Caredrivers." SilverRide sends trained escorts to provide door-to-door assistance for its customers. Both cost significantly more per mile than regular ride-hail—SilverRide's $50 flat rate for a three-mile ride is the equivalent of a 15× Uber surge fare! But these ride-hail luxe lines offer something that Uber and Lyft can't—personalized care and peace of mind.

The taxibot takeover, as industry analysts and autonomists tell it today, is inevitable. Once fares fall far enough, we'll all sell our cars and save a bundle with taxibot-powered MaaS subscriptions. But is this obsession with prices misplaced?

Consider an altogether different future, where taxibots come in a thousand flavors. Some might spend the savings of automation by staffing vehicles with more skilled crew, rather than eliminating humans from the picture. Ride attendants might serve food and drinks, provide assistive care for seniors, or help with homework. And instead of shoehorning passengers into mass-market vehicles, they can use custom AVs with interiors designed for those they serve—lowering floor heights, increasing entrances, adding handholds, and so on. Taxibots will be safer and more comfortable when they are specially designed for the disabled, the elderly, kids, or . . . leopards. One size won't fit all.

The ongoing fragmentation of the ride-hail business hasn't gone unnoticed. Big ride-hail companies are teaming up with local governments to explore the possibilities. Expanding mobility for those who can't drive or easily use transit is a high priority, and a good growth opportunity. For instance, in 2016, Boston's transit authority enlisted Uber and Lyft to supplement its door-to-door van service, allowing the city to expand and improve service. Swapping ride-for-hire for conventional paratransit vans allowed customers to order rides at a moment's notice rather than a day in advance. But the move also slashed the program's cost. Each trip made by ride-hail cost the agency less than a third of the $45 it paid for conventional dial-a-ride services. More recently, Columbus, Ohio, tapped Uber to provide disadvantaged pregnant women with no-cost rides to prenatal-care appointments and grocery stores—all in an effort to bring down one of the country's highest infant-mortality rates.

Given that some of the rich countries most rapidly adopting AVs also have very low fertility rates—Singapore and Finland, for instance—I imagine we'll soon see this idea applied there as well. When every future newborn represents such a huge social investment, it won't be long before government-provided nannybots are eagerly following expectant mothers around, ready to ferry them off to the hospital at the first sign of distress.

———

AUTONOMISTS' VERSION OF the taxibot takeover points toward a sizeable treasure. Most analysts predict that automation will be the decisive turning point in the battle for ride-hail supremacy. That's why in 2019, even as Uber and Lyft rushed to their initial public offerings, they raced to put working AVs on the streets. Meanwhile, Google sister company Waymo beat them to the punch, quietly launching a massive fleet of 600 self-driving taxis in Phoenix. With up to 20,000 taxibots scheduled to enter service in the next few years, the company could soon be serving up to a million passengers a day by self-driving cab. The day Lyft started trading, its market cap peaked just south of $25 billion. Much-bigger Uber went public a few months later just shy of $100 billion. Meanwhile, in 2018 analysts initially pegged far-smaller Waymo's worth north of $175 billion.

Or so say the simulations. Much of investors' hope for a future of ultra-cheap drone cabs rests on the surprisingly sparse findings of a handful of computer models. Carried out over the last decade by researchers at MIT, Columbia University, and the OECD, these studies used sophisticated mobility-measuring programs to deliver some spectacular results. Singapore could make do with half as many cars while still offering private rides. Lisbon, a more compact city, could swap one private self-driving cab for every six private cars—assuming it keeps its public-transit network afloat. New York City could eliminate upwards of 75 percent of its yellow cabs, but only if people share rides with strangers.

These silicon-powered studies look good on paper, but do they hold up in the real world? The first real market tests, such as Waymo's, are only just getting underway. Yet there are already serious doubts about taxibots' cost-effectiveness. One analysis, published in *Harvard Business*

Review, concluded that taxibots will cost *more* to operate than human-driven cabs, and taxibot fares could be three times higher than the cost of driving an older personal car. Not only would taxibots never achieve the 100 percent around-the-clock utilization that many of the simulations anticipated, the authors argued, but the high cost of remote human safety monitors (like those employed currently by Waymo, described in Chapter 2) would be vastly higher than currently anticipated.

Another complicating factor for the taxibot takeover is ride-hail's limited global footprint. Uber earns fully one-fifth of its revenue in only five markets: New York, Los Angeles, San Francisco, London, and São Paulo. Further growth in the places where ride-hail has worked means diversifying the services on offer. The company already offers luxury vehicles, clean-energy ones, and wheelchair-accessible rides. But automation will provide more flexibility to specialize. All of this points toward diversification for taxis in the driverless revolution, rather than a new round in ride-hail's endless race to a profitless bottom.

Uber is clearly turning in this direction. "We want to be the Amazon for transportation," CEO Dara Khosrowshahi declared in 2018, as the company moved toward forming a new kind of vertically integrated urban-mobility empire. "We are going to also offer third-party transportation services.... All of it to be real-time information, all of it to be optimized for you, and all of it to be done with a push of a button," he promised. Coming off a wave of acquisitions, investments, and partnerships that brought a vast array of bike-share, car-share, and scooter-share services under its umbrella, Khosrowshahi was well on his way to delivering on this bold vision. It is, for all intents and purposes, MaaS. But instead of haggling at an open bazaar, your experience will be more like shopping at the commissary in a company town. There'll be no apples-to-apples comparison. Instead, there will be only one kind of apple on sale—Uber's apples.

It's this possibility that raises the greatest risks. As they grasp at any hope for profitability, giant companies like Uber are moving into position to control the marketplace for mobility services. We simply cannot allow this to happen.

Microtransit Mesh

Market structure isn't the only thing autonomists get wrong in their vision of a taxibot takeover. They've flubbed their geometry too. Even if tireless taxibots do live up to the hype, and convince us all to ditch our private cars, it won't do much to reduce traffic. The opposite is more likely. There will be many more taxis than today, and they will be on the road 24 hours a day, creating traffic congestion. Even though a shift from private cars to taxibots would mean far fewer vehicles on the streets overall, each would cover many more miles per day. In one variant of the Lisbon simulation—a worst-case scenario that assumed people wouldn't share taxibot rides *and* the public-transit system was left to collapse—the total number of miles driven by automobiles on city streets nearly doubled, a catastrophic surge of traffic.

As tantalizing as the taxibot takeover may be for investors, there isn't a city on earth that can absorb that kind of surge. What we really need to keep the driverless revolution rolling in crowded cities are bigger vehicles. Urban planners have always known this, and it is why, as Silicon Valley was powering up its pipe dream of an automotive autonomous future, the European Union looked to the global South. There, the twenty-first century's answer to urban transportation was already on the streets— the minibus.

Locals call them *matatus, marshrutkas, jeepneys, angkots,* or *kombis.* Carrying 6 to 16 passengers at a time, these nimble people-haulers are the primary means of transportation for many of the megacities of Asia, Africa and Latin America. They thrive where no government-run transit exists—earning them the wonkish label *informal transit.* Yet their ridership dwarfs that of buses and trains. In Mexico City, for instance, some 30,000 *peseros* ply 1,500 different routes, carrying more than 14 million people every day—all without any central ownership or formal coordination. They move nearly 60 percent of commuters, making them one of the world's largest bus systems. Globally, the reach of informal transit is staggering—hundreds of millions of people rely on its services every day. Minivans easily carry more people worldwide than taxis and bike-share combined.

Informal transit works so well that enterprising immigrants now bring it with them. In Brooklyn, for instance, "dollar vans" ply their trade along Flatbush Avenue, plugging long-standing gaps in transit service. They charge lower fares, and run more frequently with fewer breakdowns than city buses, carrying 120,000 commuters every day. Unlike public transit, they do so entirely without subsidy, turning a profit despite paying higher driver wages and facing more regulation than back home. They're vital community hubs, too. Step aboard a dollar van and you'll find a roving mutual-support network of immigrants who share common language, culture, and occupations.

As it turns out, the appeal of this kind of commute cuts across class lines. Over the last decade, high-tech clusters have spawned startup shuttle lines almost as fast as immigrant neighborhoods have. Nowhere has the trend moved more swiftly than the San Francisco Bay Area, where the best jobs and most desirable neighborhoods for young talent are separated by a grueling 90-minute commute along one of the country's most congested corridors. Google led the way in 2007, when it began busing workers directly from San Francisco's residential neighborhoods to its Silicon Valley campus, allowing them to bypass public transit. Other big firms quickly followed suit. Ten years later, Bay Area tech companies were hauling upwards of 35,000 workers on 800 vehicles every day.

The tech giants' long-distance lines popularized the perks of private mass transit for an entire generation of California commuters. Soon, startups with names like Leap, Loup, Bridj, and Chariot scaled the idea down. These services *looked* a lot like informal transit, but, with free Wi-Fi and cold-pressed juice, a more tech-forward name was needed—*microtransit* was born. Backed by heaps of venture capital, microtransit startups cherry-picked routes, focusing on neighborhoods with concentrations of affluent, tech-savvy riders and poor transit—like Boston's Back Bay and San Francisco's Marina District.

Microtransit is hardly a new idea. In the 1970s, urban guru Christopher Alexander advised communities to create a system of minibuses "able to provide point-to-point service according to the passengers' needs, and supplemented by a computer system which guarantees minimum detours, and minimum waiting times. Make bus stops for the mini-buses every 600 feet in each direction, and equip these bus stops with a phone for dialing a bus." But back then, building an on-demand microtransit system was a difficult and costly task.

Today, the code for coordinating a few dozen shuttles and the apps to call them with is surprisingly straightforward and inexpensive to build. Better technology, however, wasn't enough to make microtransit work. Most of these first-generation microtransit startups folded—bankrupted within a few years as investors balked at the hardscrabble work the companies faced bringing costs under control. Now that they've vanished, their riders have gone back to the public-transit lines the startups once shadowed.

But microtransit will rise again. Smart shuttles hold high appeal for the ride-hail generation, especially as the value of safe screen time grows. And shuttles are already becoming just another cost of doing business for brick-and-mortar organizations of all kinds trying to lure in these distracted customers. Universities, office parks, hotels, hospitals, and even big-box stores already use them to bring students, shoppers, patients, and guests through the front door. Costs are falling—thanks to companies like San Francisco–based Ridecell, which sells web-based dispatch software complete with rider apps. And automation could slash costs further.

So what if microtransit were just another perk, offered by schools, employers, and housing developments to fill in the gaps in conventional transit? Could automated minibuses be the missing piece to making the microtransit equation add up?

The technology is ready. Are we?

———

THE GREEK CITY OF TRIKALA (pop. 81,355) was an unlikely place to put AV-powered microtransit to the test. First settled in 4900 BC, Trikala's residents were used to facing invading armies, famines, and plagues. But in the summer of 2015, a swarm of self-driving people-movers caught them by surprise.

The robot incursion was part of an EU effort called CityMobil2, which sought to build a city-friendly alternative to Google's smart car. It seemed harmless enough—Trikala would host a three-month demonstration, the project's biggest test to date. The plan was for six shuttles, snaking along a 1.5-mile route linking the city's historic quarter and its central business district.

It wasn't long before this modern technology reignited Trikala's ancient anxieties. One band of locals checked out the job-killing driverless vehicles and—already inured to the indignities of Greece's EU-imposed austerity—dubbed them "the satanic robots." A few weeks later, an AV leaped a curb, rolling a foot or two before shutting itself down (as designed). The local newspaper reported that the machine "got crazy." Then, a shadowy group calling itself The People's Front sued the city, accusing the project of causing traffic and creating safety risks for pedestrians.

Despite these ill omens, the test was a smashing success. Most residents of Trikala embraced the technology. More than 12,000 people rode the six shuttles, on nearly 1,500 trips over a three-month period. By the pilot's end, the Greeks had dispelled superstitious stereotypes with ease. Follow-up surveys showed that they were more accepting of the technology than were townspeople in France and Finland, where similar prototypes were later tested.

But Trikala's technology test almost didn't happen. In early 2014, the bureaucrats behind CityMobil2 had issued a tender for bids to build two fleets of six shuttles each. The AVs would be featured in three big demonstrations—in Trikala, the French city of La Rochelle, and the Swiss town of Saint-Sulpice. Two companies were chosen. Paris-based Induct was an auto-tech firm founded in 2004. Robosoft Technologies was a spin-off of France's leading computer science-and-automation research institute, Inria.

Each company had its strengths. Induct's was curb appeal. Its vehicle sported a sunshade top and open sides, and a rounded body with nice lines. It looked fun, like something you'd cruise around a resort in. It even had a stylish name—the Navia (Figure 4-1a). One look at Robosoft's RoboCITY Shuttle—the awkward beast that invaded Trikala—was all you needed to realize those guys had spent most of their time and energy working on the software instead (Figure 4-1b). Unfortunately, despite its better-looking design, Induct was insolvent. Just a few months after winning the bid, the company filed for bankruptcy.

Undaunted, the EU program's pencil pushers handed the work over to a runner-up, a startup calling itself EasyMile. In fact, this replacement was itself a joint venture between Robosoft and Vichy-based microcar maker Ligier Group. But the cozy arrangement proved productive, as Robosoft's technology and Ligier's knack for clean lines came together for a design makeover. The result, EasyMile's EZ10, felt more like an arrival from the future, with a broad wheelbase, rounded edges, and sleek lines (Figure 4-1c).

Meanwhile, Induct quickly reorganized and jumped back in the race to develop driverless shuttles. Under a new name, Navya, borrowed from its forebear's flagship AV, the restructured company rolled out a completely redesigned vehicle, the Arma, for a two-year test with Swiss public-transport operator PostBus. If the EZ10 was head-turning, the Arma's figure was downright sexy (Figure 4-1d).

With new vehicles in tow, the two firms set off on a worldwide race to enlist cities to showcase the new technology. In 2016 and 2017, driverless shuttles crawled along the waterfront in Perth, Australia; through the central business district of Taipei; and up and down the Las Vegas Strip—among dozens of other locations. By the end of 2017, EasyMile claimed to

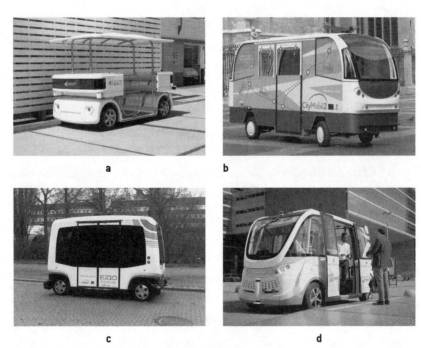

Figure 4-1. *The driverless shuttle evolves, 2011–2015.* Clockwise from top left: (a) Induct Navia, (b) Robosoft RoboCITY Shuttle, (c) EasyMile EZ10, (d) Navya Arma.

INRIA, SIGUR/SHUTTERSTOCK.COM, PER-OLOF FORSBERG, NAVYA SA.

have ferried more than 1.5 million passengers in some 20 countries. The following year, Navya sold more than 100 Armas and went public on the Euronext exchange. Both companies landed long-term financing. Navya partnered up with Keolis, the operating subsidiary of France's national railway, for a $33 million stake. Meanwhile, EasyMile raised $16.5 million from Alstom, maker of the French TGV bullet train.

While this race lacked the drama of DARPA's desert road race, thanks to the EU's efforts, driverless shuttles are now squarely on the map of self-driving's future. The quick thinking of an old bureaucracy—pulling in EasyMile after Induct's collapse—serendipitously delivered a bit of French flair to an otherwise uninspired category of AVs. And unlike the American contest, which sparked a cataclysmic wave of speculative and ruthlessly competitive investment, the EU's efforts have seeded a potentially more stable union between the French transit industry and these tech startups. The ultimate success of the driverless shuttle may well stem from this strange saga of public-sector procurement.

But the biggest legacy of CityMobil2, and the startups it spawned, is that—unlike with autonomists' vaporware dreams—millions of people all over the world rode in an AV for the first time. And in the EU's vision of the future, cities would be filled not by swift, road-hogging cars but by driverless minibuses, doodling quietly around downtowns.

DRIVERLESS SHUTTLES PROMISE most of the benefits of the taxibot takeover without the traffic and with lower emissions. But if these earth-friendly people-movers are to challenge taxibots for the urban mobility throne, they must prove that microtransit can work at scale.

The key isn't technology but real estate. Many organizations already use human-driven shuttle vans to link subprime locations to high-value crowds. Shopping centers, hotels, hospitals, universities, and condominium associations all over the world find it is often easier to shuttle their workers, customers, guests, patients, students, and residents than to relocate or expand facilities—much as Silicon Valley's tech companies discovered. Automation will only strengthen the case for shuttles. Navya claims driverless shuttles could cut operating costs by as much as 40 percent.

Government could lead the way, too. Schools could take advantage of driverless shuttles to use smaller, scattered, and more specialized facilities, since it would be cheap, safe, and effortless to move students around among them. Health-care systems could eliminate missed medical appointments by providing free transport, producing improved outcomes for patients while saving money. Better yet—they could put doctors and nurse practitioners on board shuttles to serve as primary-care mobile clinics, treating half the passengers and returning them home without ever bringing them in for costly care. Churches, temples, mosques, and community groups might roll out shuttles to tend to their own flocks, too.

But a sprawling success for driverless shuttles would create its own dilemmas.

First, who gets to ride? There's nothing at all *public* about the transit that would be provided by more *private* shuttles running to hotels, housing complexes, office parks, and even churches and schools. Shuttles are by definition exclusive. One's identity—be it employee or student, patient or parishioner—replaces the ticket as the price of entry. And as we are already wondering as we see ride-hail's expansion, will more private shared-mobility services leave public transit with less revenue to provide lifeline services for the very poor?

Second, in overwhelming numbers, shuttles could make traffic worse. While shuttles make sense in heavily car-dependent communities, they might compete with more efficient bus and rail transit in more densely populated areas. And if allowed to expand unchecked, shuttles could overwhelm roads, too. Scenes of informal transit run amok, common in the global South, should give us pause. Do we want to import the chaos of plazas and roundabouts commandeered as ad hoc depots, knotted with tangled streams of minibuses trying to squeeze through? Already, cities like San Francisco charge company buses a "curb kiss" fee, in response to locals' complaints about the unrestricted private use of public territory. They might soon complain about curbside congestion instead.

Finally—as in the days of the streetcars' expansion—how does a tangle of independent routes add up to a citywide system? Imagine trying to get around a community where private, driverless shuttles rule the roads. Even where lines cross, transfers will be impossible, since private lines won't have any reason to interconnect. Want to go from work on the east side to your kid's day care on the west side? Empty robovans will pass you by at the curb, refusing to pick you up even though they might be going your way. You'll have to go downtown first to change from work shuttle to tot shuttle, even if someone else's bus is headed straight there, because that ride is off-limits to you.

There is another possible solution, a scheme I call a *microtransit mesh*. If common sense prevails, it won't be long before a couple of companies, colleges, or condo associations see the obvious opportunity for shuttle synergy. They'll realize that two systems together are better than each alone, and combine them. A microtransit mesh would make it possible for anyone to move with ease across whatever driverless-shuttle services operated within a community. Two challenges need to be overcome to make this vision a reality—trust and coordination.

The first problem is vetting riders. In the Seattle suburb of Bellevue, a planned driverless-shuttle network funded in part by employers will soon serve a fast-growing corporate corridor where some 30,000 jobs are located. While this will help spread costs and deliver more frequent service, Company A needs to feel comfortable letting employees from Company B, or students from University C, share seats with its own people. To do this cheaply and effectively, some form of digital identity verification will be essential.

One option is to farm it out to industry. Companies like Via—which operate dozens of microtransit lines in New York, Austin, Chicago, and Washington, DC—already have rider-verification systems in place. It's easy to imagine Via licensing its tools to participating organizations or local transit agencies that want to kick-start line-sharing efforts. The challenge would be making sure there's a way to onboard disadvantaged people who don't or won't use a mobile device—and making sure everyone's data is kept private and secure.

The second challenge is coordinating those transfers, because driverless shuttles won't always run on schedules, and many won't stick to fixed routes. In informal systems, transfers are haphazard affairs, taking place at roadside shops, intersections, and traffic-clogged depots. A microtransit mesh, however, could exploit the full power of an open MaaS market to group riders by destination. Any corner or curb could become an ad hoc transfer point for two smart shuttles to rendezvous and swap passengers.

A microtransit mesh could also grow without a grand design. Participating organizations could get their feet wet with sharing, perhaps at first offering a small number of seats to outsiders, allaying fears that their own constituents would be displaced by the general public. But over time, a small success could snowball. As passengers gained access to a vastly larger network, they'd be more inclined to leave their cars behind and pile into shuttles. And the mesh would pay rewards back to operators, too, by cutting out unneeded trips and helping everyone provide better service with fewer vehicles driving fewer miles. Even bigger gains might be possible if organizations operating shuttles on different schedules shared fleets—schools and shopping malls, for instance.

Exhaust Data

Something you may have noticed about the microtransit mesh—the data flowing through it is as important as the vehicles that do. It's an ironic turnabout. For many years, the digital tailings of connected computers—like the logfiles left over after the work of routing rides and charging cards was done—were derisively known as *data exhaust.* Now, far from being a waste product, leftover data has value in predicting human behavior and is a primary driver of economic growth.

There will be plenty of data to go around. With terabytes a day per vehicle, the digital footprint of the world's AVs will soon exceed our own. By the mid-2020s, just a million self-driving vehicles could regularly produce more data than the devices of five-billion-plus connected humans.

Mobility companies are already jockeying for position to control this information—because it will be the key to understanding our travel behavior and pinpointing their pitch and pricing. But its use will go far beyond transportation, reshaping retailing and entertainment too. So much value will be created from refining the data exhaust of AVs, its collection, storage, and analysis could very well become the primary commercial purpose of the internet. But the change won't all be dollars and cents. As this information is used to choreograph our world, we'll experience a metaphysical transformation. Everything that moves will feel more coordinated, less random.

In the wrong hands, however, this power of precision control could create great risks for consumer exploitation. Data exhaust could instead live up to its name and become toxic. It will yield insights that can be used to jigger prices, tweak wages, and shuffle assets around in profitable but not always palatable ways. When data shapes the price of everything that moves, second to second, flat rates will become a thing of the past. Get ready for airline-style ticket pricing applied to road tolls, taxis, and trains. Indeed, you'll always pay a different rate than the schmuck in the lane, cab, or seat next to you. Insurers, who many thought would vanish in the driverless age, could be in for a windfall instead, shifting our perception of risk to a million different newly quantifiable dangers. Bundled by the hour with our mobility services, these a la carte insurance services will be very profitable indeed.

How fair or unfair, seamless or hopelessly fragmented, competitive or collusive these schemes become will depend on who controls the marketplace for mobility-as-a-service in the driverless future. As Uber and others advance to seize this high ground, cities are making aggressive moves of their own. A handful are building their own open versions of MaaS, as alternatives to corporate walled gardens. Since 2016, the city of Helsinki, Finland, has supported Whim, an experimental MaaS app that combines planning and ticketing for transit, taxis, bike-share, and car-share. Berlin's regional transport operator (SVG) launched its own MaaS effort in 2019, Jelbi, that integrates more than 20 different public and private transportation services. Built by Lithuanian startup Trafi, the effort draws on a highly successful deployment in Vilnius that is used by 20 percent of the Baltic capital's population. Los Angeles is pursuing a more subtle yet far more aggressive strategy. It's building a tool called RouteAPI, a sort of terrestrial equivalent of air-traffic control. This software service will dish out directions to AVs seeking permission to move through the city, according to the city's priorities for traffic control, access, and environmental impact. While it's not clear whether the City of Angels will mandate mobility providers to use RouteAPI, its very existence is a stunning power grab, an indication that officials could use MaaS to exert a surprising level of control over AV-fleet movements in the future. A more likely outcome is a future where mobility companies that opt in to RouteAPI get favorable treatment from the city's traffic signals and toll gates in return.

Despite these promising efforts, cities face huge challenges if their visions for MaaS are to prevail. For all the excitement of MaaS, and its potential to foster a diversity of mobility providers, the initial results have been modest. Subscriptions for Helsinki's Whim start at about $55 per month, and include unlimited transit use, bike rides up to 30 minutes, and discounts for taxis and car rentals. (There are pay-as-you-go options too.) But by the end of 2018, the service had just 60,000 active users, and a scant 7,000 subscribers. Close to two million trips had been booked with the app. Yet that was only one-half of one percent of the 375 million public-transit trips taken citywide that year. Berlin has been unequivocal that it wants to play the role of mobility-service integrator—rather than leaving it to Germany's powerful car companies or Uber. But Berlin's Jelbi MaaS marketplace also got off to a slow start selling subscriptions. And what's more telling, both struggled to recruit mobility companies to join their marketplace. Helsinki had the benefit of a 2018 Finnish law that effectively legislated MaaS into existence by requiring all surface-transportation providers to open up their ticketing systems. Berlin's system lacks such support and relies solely on voluntary participation by public and private mobility providers.

Meanwhile, Uber is fast expanding its own mobility marketplace. In 2019, after adding bike-share and car-share to its menu of travel modes, the company launched a partnership with transit ticketing startup Masabi in Denver, Colorado. The new service allows users to book and pay for transit through the Uber app. Notably, the service will offer side-by-side comparison, potentially highlighting the lower cost and competitive travel times of transit alternatives to taxi rides. Time will tell whether Uber continues to partner with public transit and whether it will allow competing providers in its marketplace. But it's easy to imagine this evolving into a driverless future where mobility is like streaming music today. It's just built into your phone, bundled up in your digital cloud of choice.

Not all of us will buy our own self-driving car in the future, but everyone will hire one now and then. Understanding how the shift from transportation products to mobility services could unfold, and the choices it may open or close for us in the future, is essential. Greater diversity in the taxibot takeover and the success of a microtransit mesh require that our goals for sustainability and inclusion be honored and embedded in the design of MaaS. And openness will be key. Communities need to prioritize choice, competition, and local control.

———

NOW OUR FIRST OUTING on the ghost road has come to an end. We've arrived in a new future. Driving has largely disappeared. Vehicles are more abundant, more varied, and more versatile than ever. And inside our screens—or whatever comes along to supplant them—a world of ever-expanding options surrounds us. We can click ourselves from one place to another as easily as we surf the web today.

But what have we learned? Dig a little deeper, and this world isn't quite like the one we expected.

Getting from A to B is full of choice but that means difficult decisions. We may find one answer for getting to work alone, and another for the family on a weekend excursion. A thousand variables can factor into the calculation—if we choose to let them.

Picking the right vehicle will present equally astounding options. Own or share? Silicon supervisors that coordinate the movements of entire AV fleets will let us decide. One wheel, or four? Hypervigilant code can tweak the old-fashioned physics to maintain the balancing act, if we fancy.

And once we board, we'll see computers have taken the wheel, but that just frees us up for a thousand new duties while we're on the move. Life inside AVs will be more complicated than it ever was in cars, and everyone's looking to cash in on the change. Hold on to your wallet; this vehicle is leaving the station.

Without a doubt the world that awaits us is one full of innovation. Conformity is out; specialization is in. Many will feel at home in this personalized, precision-targeted, and dynamically priced world of future mobility. The ghosts in the machine will be at their beck and call.

But what about the rest of us? What awaits us on the ghost road? We look forward to the choice. But we fear the change, the confusion, and the conflict to come.

PART II

No Man's Land

5 Continuous Delivery

For e-commerce firms, the three most important infrastructure items are information flow, cash flow, and delivery.

—Jack Ma, founder of Alibaba, 2013

My neighborhood in Jersey City was laid out in the nineteenth century, and I still shop the way people did back then. I walk to the main shopping street and call on the butcher, the baker, and the fishmonger. I live this way today because I *can*—I like the variety and the chance encounters on the sidewalk. But before household refrigeration, people *had* to shop this way. I'm told that women around here used to go out once in the morning to shop for lunch and then, after the men and children returned to work and school, again in the early afternoon to restock for dinner. It was a hard life, but also social and spontaneous. Women shopped in multigenerational groups, and the menu changed with the harvest. It all depended on what the farmers brought down from the countryside and the fishermen unloaded on the docks that day.

Few of us still live this way. In the US, almost no one does. Already by 1950, the majority of American families owned a refrigerator and a car.

They shopped less often, at bigger stores that gathered all of the shops and services of a town center under one roof—the *super*market! The fridge kept fresh meat, dairy, and produce stored safely at home. But what we gained in bulk we gave up in flexibility. When you shop like this you have to plan a week's worth of eating in one go.

We've since supersized and stretched out our shopping even more. In my lifetime (I was born in 1973), the average size of newly built houses in the US increased by more than two-thirds. "As homes grew larger, so did fridges, pantries, garages and closets," writes Alex Evans, a venture investor at Lowe's, the home-improvement giant. "The home transformed into a warehouse for storing the excesses from our retail excursions." Cue the rise of the warehouse stores—Costco and Sam's Club, both launched in 1983—selling food to consumers in bulk sizes at wholesale prices. Now you could back your ride up to the loading dock and tuck in a month-sized supply.

Today, we're on the cusp of yet another historic shift—but this time screens are replacing store shelves, and express couriers are supplanting our SUVs. Take my family. Unlike me, they're uninterested in schlepping bags down the street or hunting the aisles and shelves only to discover that what they want is sold out. Why stand in line and deal with cranky cashiers when they can consume by tap and swipe from the comfort of our living room couch? They're not alone. A decade ago, Americans left the house to shop an average of about 300 times a year. Now, it's closer to 250. By 2040 the number could fall another 30 percent, market analysts say.

Cheap delivery has removed a big barrier to the expansion of online shopping. Over half the people surveyed by UPS in 2017 said that free or discounted shipping was the main reason they shopped online. Shopping by screen is also surprisingly easier on the wallet than driving. At only $300 a year, Amazon's Prime Fresh grocery-delivery service costs far less than fetching your basic needs by car. It's almost as fast, too. Just choose your speed—two-day, one-day, or same-day.

AVs will push the envelope further still. "Driving to shop seems convenient and inexpensive, but it will be all but replaced by consumers' summoning goods to them wherever they are," predicts KPMG, a consultancy. After all the speculation about taxis and Teslas, moving goods will most likely be how we first and most forcefully feel the impact of mobility's automation.

———

THE VOLUME OF E-COMMERCE in the US has increased at an annual clip of around 15 percent for years. This steady growth is adding up. In 2018, online retailers took in more than half a trillion dollars, about one-tenth of all retail sales ($513.6 billion, or 9.7 percent, to be precise). Other countries are already further along in this retail restructuring. Just 3 percent of groceries in America are sold online, compared to 20 percent in South Korea. By the time you read this in the early 2020s, e-commerce will make up 35 percent of all consumer spending in the UK and China. With its huge base of existing stores, the US will lag a bit behind.[*]

This shift has left shop tills empty everywhere. In-store sales are still growing, but the pace is anemic in comparison, 2 to 3 percent a year. At these rates, online retail revenues will double every five years, while in-store sales will double only once per generation. Store closures continue to break records every year. In 2017, according to Cushman & Wakefield, a real estate broker, some 9,000 stores closed in the US. The firm predicted another 12,000 would shutter in 2018. Another, more recent set of estimates compiled by Coresight Research, which tracks retail technologies, found a rapid acceleration in this trend. In the first quarter of 2019, there were already as many announced store closures as in the entire preceding year. Investment bank Credit Suisse predicts that fully one-quarter of shopping malls in the US will close by 2022.

———

[*] The US is especially overbuilt with consumer infrastructure, with more than 23 square feet of retail space per person. Australia has less than half as much shop space per person, the UK and most of Europe just one-fifth.

What's more, the death of so many shops is paralleled by massive online consolidation. Rather than democratizing global trade as the web once promised, online selling is now dominated by two giants, a world historical concentration of retail power. In the US, Amazon reigns supreme, and clocked over $141 billion in product sales in 2018, a nearly 20 percent increase over the previous year. The company takes home 52 cents of every dollar spent online in its home country—more than the next 10 competitors combined—and its share is growing. With a market capitalization of close to $1 trillion, Amazon is worth more than three times its primary competitors' combined value (Sears, JCPenney, Best Buy, Macy's, Target, Kohl's, Nordstrom, and Walmart). In China, Alibaba has a similar market share, though unlike Amazon it is slowly losing ground to stiff competition.

Relentless improvement and innovation in delivery is the key to both companies' success.

Amazon's not-so-secret weapon is Prime, a $149 annual subscription service that offers free two-day shipping for over 100 million popular items. Launched in a handful of cities in 2005, by 2017 Prime had more than 100 million US subscribers. Amazon shipped some five billion items that year alone. By 2020, Prime is expected to account for half of all parcel shipments in the US. To reduce the cost and customer frustration of missed deliveries and theft, the company is aggressively pushing its reach deep into customers' internet-connected homes. Amazon Key, for instance, pairs a surveillance camera and remote-controlled door lock, allowing couriers to drop packages *inside your home* rather than risk a stolen delivery on the doorstep. Taking impulse shopping to new heights, Amazon Dash buttons let you reorder a product, such as laundry detergent, simply by tapping a small wireless widget.

In China, where few large retailers or mail-order businesses existed before the internet, e-commerce powerhouses Alibaba and JD.com have each built out their own national delivery networks from scratch. A decade ago, shipping a package the 800-mile distance from Beijing to Alibaba's headquarters in Hangzhou could take over a week. Today, the company offers overnight delivery to more than 125 Chinese cities. Both companies have exploited a fortuitous combination of cheap labor and high customer density to make the numbers work. An army of delivery workers—some *five million* strong—does the dull, often dirty and dangerous work of shuttling Alibaba's deliveries through China's crowded and car-clogged streets, moving more than $550 billion worth of goods each year to customers. During the 2018 Guanggun Jie holiday, better known as Singles' Day, Alibaba delivered *one billion* shipments nationwide in a 24-hour period. Not to be outdone, comparatively tiny JD—which has nibbled off a 15 percent share of the Chinese e-commerce market—fields a million-person courier team of its own.

Hauling all this stuff comes at a steep price. Amazon loses money each time it moves a box from its hub to your home. In 2017 alone, it spent more than $20 *billion* on shipping (from $2 to $4 per package), about 10 percent of its total take. This costly outlay, however, is buying something far more valuable in the long term—your loyalty. Among high-earning Americans, Amazon Prime's free two-day shipping is more popular than *voting*—some 80 percent of wealthy households are members.

As consumers demand faster delivery on a wider range of goods, however, Amazon is under growing pressure to bring delivery costs under control. Same-day shipping, launched in 2009 to rush big ticket items to high-margin shoppers in a half-dozen big cities, has become a new baseline. By 2019, Prime subscribers in over 10,000 US cities and towns received free same-day delivery for more than three million commonly ordered items. The share of same-day delivery is still very limited, making up 5 to 10 percent of the total parcel market in the UK, for instance. But analysts at consulting giant McKinsey forecast that by 2025 fully one-quarter of consumers will expect *all* deliveries to be same-day or *faster.*

THIS NEW AGE of instant gratification—the sudden and seamless *materialization* of online merchandise, mobilized by automated delivery—is the second big story of this book. It's a challenging prospect to make sense of. As we have seen, the way we shop has changed many times. But nothing in our past experience has prepared us for the size and suddenness of this shift. This is the stuff of *future shock*—that psychological state of "shattering stress and disorientation" that futurist Alvin Toffler described arising from "too much change in too short a period of time."

Many of us are already struggling to keep up. I'm old enough to remember life before e-commerce, before credit cards for kids, and even before package tracking. As a teenager, I'd send my money orders (gasp!) off to mail-order catalog shops. (Computer stuff. Always buying computer stuff. Some things, at least, never change.) For weeks, I'd wait for a reply to come around my corner in that brown UPS van. Then one glorious day, the doorbell would ring, the dog would bark, and with a flourish I'd put my pen to the driver's sheet and take possession. In the 1990s, online tracking arrived, and suddenly I felt like an omniscient god. Many a day started and ended watching my little bundles of joy moving through the supply chain.

Today, that sense of anticipation is gone. Instead, a nonstop stream of stuff flows seamlessly off our screens onto our streets, sidewalks, and stoops. We buy five, six, seven or more things at a sitting, only to be surprised when they arrive an hour, a day, or two days later. But delight turns quickly to confusion, as your queue of dispatched demands gets out of sequence. You ordered the self-help book *after* the family-size bag of Cool Ranch Doritos. But the snack pack got to you first. And those slippers you clicked on a week ago ... where are *they*? Presumably still out there somewhere, streaking toward you on a timetable of their own. Your world of convenience isn't under your control after all. You have a choice. You can fight it. Or you can just let go and embrace the new reality of *continuous delivery*.

This isn't just the quickening pace at which online purchases reach us, or the swelling volume either—though the average US home now receives five deliveries each week, twice as many as it did a decade ago.

Continuous delivery is what happens when you simply stop *waiting* for your orders to be fulfilled and busy yourself with the constant supply of stuff washing up on your doorstep instead. The old process of buying online vanishes, with its familiar steps—order, ship, track, receive, return. In its place, a new way of shopping moves in—subscriptions offering scheduled shipments, internet-connected buttons for one-press reordering, and in-home delivery. When everything is trackable, it's easy to treat physical goods like digital files. Swipe them off the screen and into the river of stuff headed toward your door. If you change your mind once you have something in hand, a few more gestures will take it back to where it came from in a jiffy.

Once unleashed by freight-hauling AVs, continuous delivery will have more rapid and far-reaching impacts on our communities than any future yet imagined by either autonomists or car-lite communards. Shopping malls and big-box stores will continue to go dark. Some, wrapped in mysterious-looking cocoons, will metamorphose and reopen as fully automated distribution centers. As forward-operating bases for global retail giants, they'll serve as depots for swarms of conveyors, mules, and cargo-carrying caravans that schlepp stuff into the surrounding territory for as little as four to seven cents per package per mile.

This future of continuous delivery is almost here. There's just one hurdle left. And it's a real doozy.

Conquering the Last Mile

In the twentieth century, hauling freight was something that companies did for *other companies*. We assembled trucks and built roads to move raw materials and finished goods, from fields and mines to factories, ports, and stores. But in 2018, for the first time, more than 50 percent of parcels in the US were shipped from businesses directly to consumers. By 2020 fully two-thirds will be.

This shift is already placing tough new demands on the infrastructure, vehicles, and people that move the material in our economy. Consumer goods are sent in far greater numbers of much smaller shipments, and their flow is far less steady. Fully 25 percent more packages are delivered in December than September. The busiest days see more than *twice* as many shipments as the average. The strains show everywhere—grueling hours for workers, high rents for fast-vanishing warehouse space, and growing truck traffic that crowds out all of us.

Autonomists' solution to the coming cargo crunch is, no surprise, to think big. In the world of moving freight that means one thing—tractor-trailers. Self-driving trucks, the 18-wheel kind, were one of *Technology Review*'s 10 breakthrough technologies of 2017. But before they turn our highways into ghost roads, big obstacles to automation must be overcome. Diesel trucks last for decades, and most are owned by independent drivers, who don't have the capital to buy the newest high-tech rigs. The trucking industry's past strategy for success, pushing costs onto these small-time operators while demanding better results, has left it ill-prepared to invest in the vehicles of the future.

But AVs will make a decisive difference—and soon—by conquering cost in that hellish territory shippers call *the last mile*. Let's be clear, though. While this phrase adds a modern businesslike punch that the traditional term *drayage* lacks, it isn't used with affection. Its origins date to the final moments of the oldest test of human endurance, the marathon. As legend has it, at the 1908 London Olympic Games, an extra 385 yards were added to the 26-mile marathon course in order to put the finish line directly in front of the royal family's viewing box. As they staggered down the final stretch—the dreaded "last mile"—exhausted racers mustered what energy they could to shout, "God save the Queen," honoring the monarch as they gasped for air.

Modern use of the term is far more mundane, but reflects a similar weariness. During the telecommunications boom of the 1990s, *the last mile* became synonymous with the headaches of running wires from central switches to homes, a capital-intensive, low-margin business. The job was—and still is—a thankless, never-ending battle against backhoes, weather, vandals, and pests. For continuous delivery, the trials of the last mile are even greater. One must overcome the tyranny of geography not just once—to pull a single fragile wire into place—but every hour of every day.[*]

The campaign begins at regional *fulfillment centers*, the massive warehouses whose collective footprint is growing in lockstep with the boom in online shopping. In the UK, for instance, for every additional £1 billion spent online, about 800,000 square feet of new order-fulfillment space is needed. These buildings are the server farms and master switches of this internet of stuff, the material manifestation of online shops. Containers from far-off fabricators are offloaded, and the process of logistical fission begins. "Large shipments of goods atomize into hundreds or thousands of individual deliveries, each with its own route, location, and timing." A vanishingly small number of people are involved, as much of the work inside is itself heavily automated. Amazon's newest fulfillment centers are the size of small towns, covering more than one million square feet, yet employing fewer than 1,300 people on average. Instead of using human labor, these behemoths are brought to life by cheap electricity, high-capacity roads, and robots.

Still, they aren't nearly enough to slake our thirst for shipped stuff. Continuous delivery's short windows—anywhere from 48 hours to just 4—require the deployment of miniature replicas of these massive plants closer in to the urban core. In big cities now you can find them in a circle around the downtown like planetoids orbiting a big star, precisely six to nine miles out from the geographic center of population.

From here, the heavy lifting is still done by hand, a costly endeavor. Same-day delivery in big cities has historically relied on an itinerant work-

---

[*] The use of *last mile* in shipping is analogous to its use in passenger transportation seen in Chapter 4, where it described the final segment of a journey after one has exited the transit network. In both cases, the term is used figuratively. The actual distance can range from a few yards to several miles.

force of independent contractors. This setup allows vendors to easily scale the pool of workers with demand.

Today, a new courier workforce is on the move, as professional delivery drivers are replaced by phone-guided gig workers dispatched by apps like Uber Eats and Deliveroo. These gadgets allow companies to hire less-skilled workers, but in the switch, valuable local knowledge and curbside experience have been lost. For instance, one UK study identified a veteran local driver who logged 44 percent fewer miles in 35 percent less time than a novice making the same set of deliveries. The deskilling of local delivery has also inevitably driven down wages, provoking a growing worker backlash against firms like "Slaveroo." The centralization of dispatch to the cloud has also eliminated the local middlemen who once provided a handy pressure point for police when traffic, parking, or road-safety issues came up.

For shippers, however, the costly gains obtained by deskilling delivery haven't been enough. With few remaining options, many are turning to automation to slash costs and increase speed over the last mile. Three opportunities on the horizon may allow shippers to fully reconcile the ledger of the last mile.

$$\equiv$$

THE FIRST OPPORTUNITY is finding a better fit between human and machine. Today, almost all parcel deliveries travel the last mile by truck or van. Food may go by cycle or scooter but often moves by car—and it still requires a person to be involved. In the future, delivery AVs will be easier to right-size for the job. Goods-hauling AVs will offer a wide variety of shapes, sizes, and speeds for moving stuff. And to a far greater degree than their passenger-carrying cousins they'll be able to dispense entirely with a human crew.

Conveyors, those diminutive sidewalk-roaming machines we met earlier, are great for moving small, perishable items over short distances, adding pennies to the price of a meal. Starship Technologies, the maker of the conveyor I met in Tampa (Chapter 3), claims it can slash the cost of delivery by 90 to 95 percent from current levels. In 2018, the company launched a pilot at the Silicon Valley campus of software giant Intuit in Mountain View. There, six conveyors ferry breakfast, lunch, and coffee from the company cafeteria to hungry employees along a network of pedestrian paths. Demand is highest at breakfast, because deskside delivery allows workers to skip straight from the parking lot to their office without stopping at the canteen.

Mules are closer to the size of minivans but still svelte compared to conventional delivery trucks, and they will move to the beat of a different drummer. Cody, a mule concept promoted by design firm IDEO in 2015, is tailored for deployment into busy neighborhoods. A futuristic cross between a plastic storage box and a minivan, this urban workhorse is designed to ferry an entire day's worth of deliveries to a convenient corner location, serving a whole city block's worth of customers in one trip. Mules don't just deliver, however, but also take returns—all while replacing a swarm of dirty, noisy, and often dangerous delivery trucks. Automakers have a half dozen mule prototypes already in production. Toyota's mule, the e-Palette, lists Amazon, Pizza Hut, and Uber among its partners. Another Palo Alto–based startup, Nuro, developed its own mule from the ground up in less than two years, and raised $940 million in venture funding in 2018 to roll it out.

Looking further out, there's hardly an idea for moving stuff that isn't being explored as an AV. At least three different companies sell semi-autonomous suitcases that can follow you through the airport while avoiding obstacles like people, pillars, and walls. The Puppy 1, made in China, even balances itself on two wheels. Mercedes sees a future for its Sprinter delivery vans to serve as mother ships, equipped with rapid-deploy ramps for up to eight Starship conveyors. One can only imagine the impression this blitzkrieg of delivery droids will leave on the neighborhoods it serves.

The design dreams don't stop there. Freight AVs will push the edge on automated driving, too. Cargo can't get carsick and doesn't have to survive crash tests. It can't complain about aggressive driving, and won't mind a painfully cautious pace either. Free of the need to keep passengers calm or unscathed, AV haulers can experiment with high-G acceleration and deceleration, extremely fast or slow travel, and new means of loco-motion beyond wheels and tracks. Since no one is inside to bang around, they can be programmed to deploy hard emergency braking that would otherwise injure passengers or to sacrifice themselves (and their cargo) to prevent harm to others. Operating at night in industrial districts, and at other venues where few humans are present, they'll present less risk to people on the street.

Investors will fuel innovation in freight AVs, too, as they search for faster payback than they'll find from passenger AVs. The market need is immediate—online retailers desperately want to cut delivery costs as they expand into low-margin businesses like groceries. And freight AVs will be easier to get to market than passenger types. They'll be simpler to develop, since they need only to learn the ins and outs of a small territory, and can even be confined to sidewalks or sparsely traveled streets. That means much smaller maps and simpler software are needed than for a driverless passenger car. And they'll be less regulated than anything that carries people, reducing the time and cost of testing.

LAST-MILE OPPORTUNITY NUMBER TWO is the wasted cargo-carrying capacity rolling by us, right now, in the trunks of taxis, back seats of cars, and empty luggage racks of buses. Like ride-sharing, piggyback-ing deliveries on vehicles headed past your door is an idea that's already being put to use in human-driven vehicles. Atlanta-based Roadie, for instance, brokers shipments of bulky, slightly less time-sensitive items such as lost airline luggage and home furnishings with drivers taking long commutes or road trips. The company works with major airlines and retailers like Home Depot and Macy's, and claims to have a footprint larger than Amazon Prime, having serviced some 11,000 communities in its short history.

AVs unlock lots of new possibilities for this kind of peer-to-peer parcel shell game. Instead of paying to park all day, why not rent out your self-driving car to deliver groceries while you work? Or how about having your stuff delivered to your car, wherever you are, instead of to your house? That's Amazon's thinking, which it hopes to develop into a cheaper, better, and bigger alternative to its network of parcel lockers, which have had high costs and lackluster appeal. The company's Key In-Car service serves 37 US metro areas and will test whether people feel differently when the locker is their own car. Key In-Car allows customers to receive secure, unattended trunk deliveries to a variety of GM and Volvo vehicles at no extra charge. For now, the deliveries are made by gig workers. But how long will it be before we stumble upon AV mules and self-driving SUVs "mating" in the parking lot, as a clandestine swap of packages passes between them?

Still to come are even more ambitious schemes to tap the dark tonnage lying idle in transportation networks. Wasted cargo-carrying capacity abounds in the millions of commercial and government vehicles that sit unused much of the day and night. That's the insight behind Hannah, a self-driving school bus proposed by Seattle-based design studio Teague. The six-seat student jitneys would be the backbone of an on-demand school-shuttle system in the mornings and afternoons, picking students up as an Uber would. But in between and overnight, they could make deliveries. With nearly a half-million school buses in the US, such a move could double current last-mile delivery capacity. By comparison, one of the three major American parcel-haulers—UPS—rolls only about 100,000 vehicles on any given day. As a bonus, AV school shuttles doing double duty as delivery drones would provide desperately needed revenue for underfunded public schools.

———

LAST-MILE HACK NUMBER THREE exploits automation's unique ability to work around the clock at no added cost. Nighttime operation has long made sense for shippers. It frees them to move goods quickly and cheaply through densely populated areas when roads are clear. But the high cost of overnight wages and the loud noise of heavy trucks limit the use of night delivery.

Both problems will all but vanish with changes in vehicle technology. Electrification will dramatically reduce the noise from commercial traffic. In fact, electric vehicles (EVs) have a long history of use during predawn hours in British cities and towns. More than 25,000 "milk floats," tiny delivery carts with open sides for easy access to bottle racks, were commonly found trundling through towns and villages from the 1930s into the early 1980s. As *The Economist* notes, "Short ranges and low top speed were unimportant for a milk round but near-silent running meant customers could sleep." Labor costs won't be higher overnight, because robot drivers don't need to be paid a premium to work the graveyard shift. What's more, provided that the tail end of delivery can also be automated— loading bays, stockrooms, and refrigerators for perishables—overnight delivery can actually reduce costs for receivers, too.

Nighttime delivery is even an attractive option for shippers who don't intend to eliminate any workers. Less traffic at night means shippers can use bigger vehicles to make more deliveries, and bulking up reduces per-unit delivery costs significantly.

Oddly, AVs designed for the night shift might actually be put to higher-value use during the day, as battery backup for a wind- and sun-powered grid. An unsolved obstacle to the widespread use of renewables to charge electric vehicles is that we will typically need to charge them overnight, after the sun has set and the winds are still. To provide the needed power to recharge millions of EVs, future electric utilities will need a way to store power produced by renewable sources during the day. Such storage is expensive, and it will take many years to build—potentially decades if new storage technologies don't pan out.

Electric AVs scheduled for the night shift, however, make the perfect counterpoise for consumer EVs. During the day, when windmills and solar farms are at peak production, night mules would sit parked at charging points, their batteries soaking up power that might otherwise be lost. If demand for power spikes, utilities could draw that juice back down instead of turning to the dirty gas turbines they'd typically fire up to fill the shortfall, avoiding extra carbon emissions.

At least that's the opinion of Henning Heppner, founder of Berlin-based EV-charging pioneer Ebee, who knows the territory well. By his estimation, it could be more lucrative to rent out night mules as energy storage than to use them for their intended purpose as package couriers. Given the huge revenue potential, which could rival their value as parcel-haulers, this electricity juggling could easily become the fleet's main business purpose.

Heppner laid out this vision during a 2018 visit to EUREF, a high-tech business campus in Berlin's Schönberg district where his network of startups is prototyping a wide range of urban technologies. My mind raced as I took his forecast at face value. I imagined a future city, its streets swarmed with conveyors and mules ferrying every object under the sun. But should that sun disappear and the first algorithmic signs of distress ripple out from the electrical grid, they'd suddenly disengage from delivery duty, return to the depot, and dump their remaining joules back into the faltering grid. It's an enticing vision, one that could clinch the case for full conversion to an all-renewable grid powering an all-electric fleet.

Just don't expect your deliveries at the end of a cloudy, windless day. The trucks will be dead from doing double duty as your neighborhood's battery.

When Atoms Move Like Bits

The battle to conquer last-mile distribution, waged by robot armies, will be one of the great business stories of the next decade. It will certainly be won. In the twentieth century, the cost of moving freight declined by 90 percent, according to a study published in 2004 by Harvard University economists Edward Glaeser and Janet Kohlhase, who assembled a mountain of historical data on American manufacturers' shipping costs. "There is little reason to doubt that this decline will continue," the Harvard professors concluded.

But how far down can shipping costs fall? The three last-mile innovations we explored (unmanned AVs, package lockers, and night delivery) each promise huge cost savings, according to analysts—50 percent, 35 percent, and 40 percent, respectively. Obviously, there's some overlap, since these numbers add up to more than 100 percent. But it is clear that, deployed in combination, the gains could be huge. Could the driverless revolution shave *another* 90 percent off the cost of moving goods? And if so, what impacts would such near-zero-cost shipping have on shopping, stores, and streets?

A good starting point for exploration is a simple graph showing the relationship between supply and demand for the shipment of parcels (Figure 5-1). Don't worry if you slept through microeconomics; it's pretty straightforward. At left, on the vertical scale (or y axis), the price of delivery increases as we go up. Along the bottom, on the horizontal scale (the x axis), as we move from left to right the total amount of shipping consumed increases. The two solid curves show the world as we know it. The supply curve (S1) rises up and to the right, indicating the degree to which shipping companies like UPS, FedEx, and the postal service are willing and able to increase capacity as prices rise. Demand (D) responds in the opposite way—when the price falls, consumption increases. Where the two curves meet, the market reaches equilibrium (E1), and the prevailing price (C1) and level of supply (F1) are set.

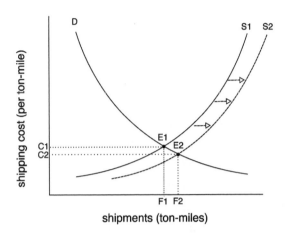

Figure 5-1. *The effect of falling costs on demand for shipping.*

Now, consider what happens when a new technology like AVs is introduced. Let's say one day Robohaulers, Inc., enters the market with a million-conveyor fleet and with this last-mile solution can move the goods at 25 percent below what other parcel-haulers charge. The result is predictable. Robohaulers will undercut the competition, who must match the new price, and the price of shipping falls (to C2). But now that prices are lower, consumers respond by buying more, and the supply curve slides to the right (the dashed curve S2). The market settles in at a new equilibrium (E2), where costs are lower but freight traffic is higher. This isn't rocket science. It's capitalism.

Like much economic theory, this graphic is a deceptive simplification of the real world. The market's adjustment to last-mile automation will be a bumpy ride, as these technology-induced surges in demand become both more sudden and more frequent. Each innovation—whether it's night delivery or right-sized vehicles—will trigger an abrupt increase in the volume of goods moving along our streets. The impacts will be disproportionately severe in cities and metropolitan areas, where roads are already close to capacity. Meanwhile, the total amount of shipping will continue to swell, along with the carbon footprint of making and moving all those goods.

Thankfully, at some point the pace of disruption will peter out. There will always be some sources of friction that impede the free flow of goods. These include the cost of fulfillment centers themselves (Amazon has over 100 in North America that cost $100 million each to build and consume enormous amounts of electricity); maintenance, repair, and remote monitoring for millions of conveyors and mules; and the lingering challenge of the *last meter*, where gates, stairs, and elevators stymie even the smartest robots. What's more, eventually the law of diminishing marginal returns kicks in. Just as that eighth sundae you eat in a single week is less valuable than the seventh, which fully satisfied your daily craving for ice cream, at some point consuming more produces less value for the consumer. (Though as we'll see in the next chapter, cheap shipping will also encourage businesses to do more of the work of producing goods and services using shipping themselves.)

I also find some mild comfort in the fact that we human beings will provide a modicum of challenges to retard the march of continuous delivery. Synchronization with fickle and unreliable humans will always create overhead. "People are never ready on time," says David King, a professor of transportation at Arizona State University who studies AVs. "They're not waiting for a package to arrive. They're going to be on the phone, or in a coffee shop." Even with automation, for these systems to work well, he argues, the recipient has to be ready.

A more substantial dampener on the growth of shipping is the likelihood that most of shippers' reduced costs won't be fully passed on to consumers, or at all. With shipping already free, all of Amazon's current efforts to control costs—such as its "Amazon Day" promotion that consolidates Prime subscribers' deliveries to one day a week—will go directly to its own bottom line. And even as producer costs fall, consumer prices for delivery could *rise*. In 2016, McKinsey surveyed nearly 5,000 consumers in China, Germany, and the US and found that more than one quarter (in each country) were willing to pay "significant premiums for the privilege of same-day or instant delivery." Since it is most often impatient, high-spending young shoppers who choose speedy delivery, the take from such surcharges could be a big business.

———

ALL THINGS CONSIDERED, a decade from now, sending atoms across town will be *almost* as cheap and effortless as sending bits across the globe is today. Not free but, say, 90 percent cheaper than today. How will we exploit this new material mobility? And what larger changes will it unleash?

Oddly, our best glimpse to date of a world of near-frictionless freight has come and gone. Few of us noticed. From 2013 to 2018, San Francisco–based Shyp offered an on-demand, no-pack shipping app that turned your iPhone into a *Star Trek*–like matter transporter. If you wanted to send something across the city, or the country, all you had to do was point your mobile at an object, click a pic, and—zoom!—off it went. Well, sort of. You still had to wait for a courier to roll by and whisk your stuff away. But there was no packing, no waybills to fill out, and no hassle. For a flat fee of five dollars (on top of the actual shipping cost), Shyp boxed the stuff up for the journey and handed it off to a long-distance hauler. A few days later, up popped a notification on your screen. Your stuff had arrived in Austin, London, Tegucigalpa, or wherever you'd beamed it to.

Shyp was *way* ahead of its time. When the service launched, most of us were still getting used to sending selfies. The idea of transmitting stuff by taking a picture and clicking Send was still too futuristic to comprehend. After a bumpy five-year run, the company imploded, before anyone fully grasped its significance. But it is an idea that inevitably will come back around, and AVs will be essential to making it work.

More than a decade after the DARPA Grand Challenges and the much-ballyhooed launch of the Google Car, the idea of moving people by self-driving cars still feels iffy, like speculation of a far-off world we might just as well live without. But the technology and the rationale for driverless movement of goods are quickly taking shape. One day soon, I suspect, the last few pieces will fall into place, and sending things will suddenly become so cheap that—like sending texts—we simply don't think about the cost anymore.

And that, more than taxibots and Teslas, will change the way everything works.

6 Creative Destruction

What we're trying to do at Pizza Hut is get the ovens closer to the front door.

—Artie Starrs, CEO, Pizza Hut

Life in Manhattan is a series of moments that take place inside boxes. You live in a box called an apartment, and (for many of us) you work in another box called an office. That box is itself subdivided into cubicles, which are tiny boxes themselves. You travel between home and office on the subway, a long chain of rolling boxes. We've even caged nature in with right angles. Central Park is, after all, the city's biggest box. Spend a bit of time in the Big Apple, and the inexhaustible supply of walls can close in on you.

A low-altitude flyover is a good way to find a brief escape and provides a new perspective on the city's relentless rectilinearity. Any number of helicopter-tour operators are happy to oblige. At fifteen hundred feet, a larger logic dominates your field of view. You realize that the street grid's structure, laid down in 1811, doesn't just box us in but also connects us out. Wide north-south avenues were intended to unlock land development by expediting the flow of commuters. They succeeded, all too well, and have been clogged with traffic ever since. But it was the profusion of east-west streets that linked us to the seas, by making it easy to move goods inland from two mighty rivers—the East and the Hudson. Broadway gets all the attention, but crosstown connections made New York the world's biggest and most efficient industrial trading hub. Nowhere else could so much freight move from ship to loft and back again so quickly and cheaply. The freighters are long gone, but the grid remains. Now it facilitates the flow of Ubers, buses, and bikes—and an army of trucks and vans delivering more than 1.5 million packages each day, triple the number a decade ago.

Just a few miles to the west, but a world away, another flight is lifting off. Weaving between the rafters of a massive fulfillment center in New Jersey's Hackensack Meadowlands, a drone takes in a sweeping view of the vast floor below, and a different sort of street grid emerges. From this vantage point, the matrix of stacked crates and the hundreds of conveyors trundling among them mirror the low-rise industrial sprawl that reaches for miles in every direction beyond the building's four walls. Closer examination reveals that it's the stacks themselves that are on the move, rolling along on robotic sleds.

This new cargo cult is serious business. In 2012, Amazon cornered the market on these warehouse AVs, acquiring Massachusetts-based Kiva Systems for $775 million. Each year since, the megaretailer has added 15,000 droids to its global robot army, with stunning results. The company has cut the time from when you place an order to when it rolls onto a truck ("click to ship," in the biz) from over an hour to just 15 minutes, while reducing costs by 20 percent. By early 2019, over 100,000 of these diligent devices were reporting for work 24 hours a day, 365 days a year, at more than 100 Amazon distribution centers.

Getting rid of humans clears the way for a more dense and intensely kinetic industrial architecture. Thanks to automation, despite covering footprints of a million square feet or more, these warehouses of the future are downright claustrophobic. The distance between shelves is narrowed, cramming 50 percent more inventory into the same square footage. AVs rocket among the rows at speeds of up to 17 miles per hour, more than four times a brisk walking pace. Workers are confined to safe zones.

Even more alien interiors are to come. Boxed out of borrowing Kiva's patented sled scheme, Amazon's rivals are pursuing their own exotic ideas for robot workers. Ocado, an online supermarket chain in the UK, has bots that slide along tracks suspended above bins of goods, grabbing up items as they traverse an illuminated grid reminiscent of the 1982 film *Tron*. China's JD.com employs a freakish spiderlike robot that hangs from the ceiling swinging its five-foot arms around more than 100 times each minute to pick and pack items for shipment.

These insectile swarms of intelligent machines infesting their cardboard colonies are dreadful, indeed. But they are a necessary evil. This vast machinery has a job to do, sustaining a human population whose material demands are soaring. If present trends continue, by midcentury, there will be more than 6.5 billion city dwellers worldwide—half again as many as the four billion plus who call the metropolis home today. They'll be wealthier, older, more educated, and far more demanding, clamoring each day for their books, toys, meals, and garments. Yet they will face excruciating pressure to control carbon emissions. These are the machines being mobilized to feed, clothe, and care for us during the complicated trade-offs of the century to come. We simply won't survive without them.

This "city" of mechanical minions is, for now, walled in. But can this ruthless geometry be contained? Or is Amazon's inevitable next move to carry this new town plan out into the larger world? Rarely has capitalism produced such a pure mechanism for "creative destruction," that "process of industrial mutation" described by Joseph Schumpeter "that incessantly revolutionizes the economic structure from within, incessantly destroying the old one, incessantly creating a new one." Will these parcel-purveying machines prove as ruthlessly efficient at restructuring our communities, too? And if they succeed, will there will be room for us in this new city of boxes?

Because right now, it's obvious that they'll do their best work only when *we* get out of the way.

———

THE TYPICAL DOMINO'S PIZZA outlet averages a mere 1,500 square feet, and has all the ambience of a big-city bus station. But for better or worse, it is the working model for the restaurants of the driverless future. Before Amazon pioneered its dirt-cheap, no-frills Basics brands—brought to you with free same-day shipping—the folks at Domino's had long since discovered that customers preferred speed over quality. In 1973, the pizza giant began its famous "30 minutes or it's free" campaign. The guarantee went away in 1993, but not before forever branding the company's demanding delivery standard in our collective unconscious.

Of course Domino's didn't invent pizza delivery, but it did expand this practice to an industrial scale. Human beings today eat more than five billion pizzas every year. Domino's alone delivers one million per day. The cardboard pizza, scant sauce, and synthetic-tasting cheese (replaced with real natural cheese in 2008) helped keep costs down, for sure. But it was the out-of-the-way, hole-in-the-wall storefronts, coupled with telecommunications technology, that really made the difference.

Drop in on one of these forlorn fronts for the pizza syndicate between shifts, and you'll find the meager staff milling about. There's little here but a kitchen, a cash register that doubles as a dispatch desk, and a few seats that serve mostly as a drivers' lounge. Make a call, click a mouse, or tap a screen, however, and the shop springs to life, sending a hot pizza streaking out into the darkness of night.

Domino's decision to favor such low-rent surroundings was the key to a simple winning strategy—swap cheap transportation for costly real estate. Instead of paying high rent for Main Street shops with the greatest foot traffic, franchisees could choose any out-of-the-way strip mall or industrial park with good road access and a bank of phones. And by outsourcing their dining room to ours, Domino's changed the entire cost structure of the pizza business.

In the digital age, the model that Domino's pioneered is spreading from fast food to an ever-expanding variety of cuisine. These gourmands-on-the-go are known as "ghost restaurants"—simple slopshops that dish out cheap food to impatient clickers on portals like Grubhub and Seamless. What's driving the shift is a combination of digital dynamism—smartphones are proving just as useful in overhauling delivery dispatch as they were for taxis—and the intense culinary competition for a younger, more affluent, and more informed crowd.

Like the pizza pioneers, these virtual bistros radically slash costs and streamline service. Green Summit uses kitchens as small as 2,000 square feet. When the company opened in Chicago in 2017, it packed nine ghost restaurants offering Chinese, Mexican, sandwiches, and more into a single location. Kitchen United, another startup, banked a $10 million investment from Google's venture arm in 2019. It plans to open 400 locations offering up to 20 different menus apiece. Both chains offer tiny dine-in areas—useful for chefs to entertain investors and get face-to-face feedback from customers—but by reorienting almost entirely toward delivery and take-out they reduce overhead while offering an alternative to chains like Pret a Manger. Green Summit needs just one-quarter of the floor space but claims it can serve 90 percent of the sandwich giant's clientele, who simply grab a meal and go.

The disruptive economics of ghost restaurants is driving an explosion of culinary creativity. Their ultralow startup costs make it easier for younger and more daring chefs to jump into the food business (much as food trucks have). These ultralight eateries are more flexible too. Programmable signage means lower costs for menu changes, which allows ghost restaurants to try new concepts with lower stakes. Green Summit's founder says the company loses as little as $25,000 if a new menu bombs.

Traditional restaurants could easily spend 10 times that amount or more to try out a new trend like poke.

As it grows to industrial scale, the sweeping dematerialization of sit-down dining is spawning its own infrastructure. Between 2012 and 2017, more than $10 billion flowed into last-mile food-delivery start-ups, like Postmates. But the most interesting action is at the back of the house. While Domino's dominated the pizza business from suburban strip malls, a group led by Uber founder and ex-CEO Travis Kalanick is instead snatching up parking garages around the urban core and converting them to shared kitchens and delivery depots. In 2018, the group took control of four in New York City alone, at a cost of more than $40 million. It's said that if you want to get rich in a gold rush, you should sell picks and shovels. Taking a page right out of the prospector's playbook, Kalanick's CloudKitchens, as it's called, will rent aspiring young chefs their first canteen instead.

—

DOWN THE HILL from my Main Street, in the neighboring town of Secaucus, a cluster of facilities has in recent years become ground zero for the retail liquidation to come. This mishmash of tidal wetlands and asphalt industrial parks, linked to Manhattan by two vehicular tunnels, the Northeast Corridor rail line, and the world's busiest bridge, has become a go-to spot for same-day distribution into Manhattan. With (relatively) cheap land and great transportation, surrounded by tens of millions of consumers, it is an ideal place in which to build out the push-button, automated shopping system of tomorrow.

Pick any building at random here, and you'll get a sneak peek at the future of continuous delivery. Inside one windowless, brown 160,000-square-foot box is Rent the Runway, which promises "a dream closet on your terms"—essentially a subscription service for high-end women's clothing delivered by courier. Think of it as a ghost department store, stocked to the gills with designer brands that are carried by couriers to customers who want to dress to impress—at parties, job interviews, or any other occasion. When this facility opened in 2014, it was the nation's largest dry-cleaning operation. Today, it's dwarfed only by the company's second plant, in Arlington, Texas, which is twice the original's size.

Another building houses MakeSpace, which works just like the cloud file-storage service Dropbox, but for your stuff. "Never visit a storage unit again," the company's pitch goes. Call them up, and they'll come take your belongings away for safekeeping. You can even snap photos of your crates' contents with their app, browse them online later, and submit a request for delivery, all from the palm of your hand. I wasn't kidding about the city of boxes. This is a service for space-constrained box-dwellers to stow away their extra boxes inside other boxes somewhere else.

Hop on the New Jersey Turnpike and cruise a few miles down the road to Blue Apron's half-million-square-foot plant in the town of Linden. Here, a smallish army of food-prep workers is busy preparing meal kits—fresh, prepackaged ingredients, spices, and recipes—for shipment to some of the company's 750,000-plus subscribers nationwide. Blue Apron has one-upped the ghost kitchens. Not only do you get to host this meal, you do the cooking too.

The consolidation these companies achieve is remarkable. By swapping programmable mobility for proximity, they substitute the inexpensive labor of New Jersey drivers for the astronomical rents of Manhattan storefronts. Activities that were scattered across an array of neighborhood shops, storerooms, and kitchens are concentrated under a single massive roof, generating enormous economies of scale. Being close to your customers is no longer a function of location alone but mainly how well your dispatching software keeps your fleet clear of traffic jams and on schedule.

The exodus of low-margin commerce from high-cost centers is only beginning. Designer clothes, self-storage, and gourmet meal kits are merely the first lines of business to ghost Main Street. Soon, as AVs drive the cost of delivery toward zero, the process will speed up. Pizza may once again lead the way. In 2018, Pizza Hut partnered with Toyota to turn the automaker's e-Palette mule concept into a roving food factory. A year later, Domino's teamed up with mule manufacturer Nuro to test pizza delivery in Houston, Texas. In a few years, AVs carrying an entire kitchen, with hot and cold storage, ovens, and a pizza-making robot, will roam the roads near you.

But how far could this robot-powered restructuring of dining, retail, and neighborhood services go? It's easy to imagine an armada of mules rolling in, autonomous pop-up peddlers selling not just pizza but petunias and puppies, too. Algorithmically set for aggressive pricing, they'd follow sudden shifts in demand, hollowing out retail districts as they went. And short of putting up physical or legal barriers, no jurisdiction could fence them in.

From Kipple to Circular Economies

Every Tuesday night, I go to my second job. I'm the superintendent of a small building—my house. There's only one family that lives here, but it produces a steady and surprisingly vast stream of boxes that must be broken down for recycling. It's satisfying work. There is a catharsis to sliding my box cutter open and slashing away, reducing the mountain of refuse to a neatly bound pile of recoverable material.

Over the last year the cardboard accumulation has gotten out of hand. And I'm not the only one drowning in the stuff. Waste Management, which hauls trash and recyclables from more than 20 million homes in the US and Canada, reports a 20 percent increase in curbside cardboard collection in the last decade. Wrestling ever-larger piles to the curb, I wonder what all these containers originally carried across my doorstep. I dread the thought of spring cleaning, when last year's must-haves become this year's clutter—and none of that junk can be recycled. Philip K. Dick, the science fiction writer, had a name for this sedimentary accumulation of consumer artifacts. "Kipple," he wrote in the 1968 novel *Do Androids*

Dream of Electric Sheep?, "is useless objects, like junk mail or match folders after you use the last match or gum wrappers or yesterday's [newspaper]." Today's throwaways are far more functional than Dick ever imagined. Single-use packaging has given way to all-but-disposable products like coffee pods, iPods, and assemble-yourself furniture. But the mound of matter is every bit as monumental as he predicted.

It's clear that even in today's embryonic form, continuous delivery is overwhelming us. But as shipping costs collapse, the driverless revolution will unleash a new wave of kippleization. Will junk mail give way to junk stuff? Instead of pop-ups on our screens, will we face pop-ups in real life, as e-tailers prospectively send potential purchases without permission, in the mere hope we won't swipe them away? As I gawk at my mound of kipple, imagining a future of spam that materializes out of thin air, I fear the worst is yet to come. "When nobody's around, kipple reproduces itself. . . . The entire universe is moving toward a final state of total, absolute kippleization," Dick predicted.

MY GROWING FEAR of kippleization is underwritten by resource economics. It may be true that the stuff that's delivered to my house today is simply replacing stuff we would have bought at the store, but as shipping becomes the cheapest way to do *everything*, it won't cannibalize retail alone. It will become a more widely used ingredient in the way all products are made and every service is delivered.

To understand why, we need to travel back to the high days of the Industrial Revolution, when British factories ran on coal. For decades, the empire chugged along, burning through an abundant natural bequest of fossilized deposits. By the 1860s, however, it began to dawn on the ruling class that the mines would soon be tapped out. Improving the coal-burning efficiency of furnaces became an urgent priority. But would it stave off the inevitable exhaustion of this all-too-limited resource? This was the subject of an inquiry by William Stanley Jevons, a young economist, in *The Coal Question*, published in 1865. The answer—a decisive *no*—has confounded conservationists ever since. Because what Jevons argued was that improving the efficiency of coal furnaces would lead to vastly *more* coal consumption, rather than less.

Jevons reached this counterintuitive conclusion via rational inductions. Let's say one day a steelmaking plant installs a new type of coal furnace that's significantly more efficient than existing technology. Because less fuel is needed per unit of steel produced, the company's profits will immediately increase. Now the firm's owners face a choice: Keep producing the same amount of steel or raise their output? Conservationists would urge them to stand pat, thus lowering their overall coal consumption (and pocketing the extra profit). But capitalists would see an opportunity for growth. They'd urge the steelmaker to slash prices, expand production, take market share from their competitors, and earn larger profits overall (even though per-unit profits would shrink). As Jevons realized, this would force competitors to do the same, by getting hold of their own coal-conserving technology, or be forced out of business. And here's the painful irony of what's come to be called the Jevons paradox, or the *rebound effect*—even though our hypothetical steelmaking sector has converted (quite swiftly, in fact) to the new coal-conserving technology without any government subsidy, it is using *more coal than ever*. The invisible hand of the market giveth, and it taketh away.

Things only get worse from here, as the effects of cheap steel made with efficient coal-produced power ripple through the economy. Now that steel is being overproduced, its price falls relative to other materials. Cotton, wood, and other labor-intensive raw materials whose production didn't benefit from the leap forward in coaltech are set aside as other industries develop new ways to exploit the abundance of cheap steel. Steel houses are all the rage, and the middle class develops a new taste for steel tableware. (Perhaps intrepid fashionistas start wearing steel pants, too.) The sudden upsurge in steel demand only encourages steelmakers to pour their profits back into coaltech research in the hopes of gaining an edge in the boom. The whole cycle starts over again, and coal consumption rockets ever skyward.

The conservationists' takeaway from Jevons's thought experiment is both clear and demoralizing—in a free-market economy, efficiency improvements alone can't reduce resource consumption, and are likely to increase it instead. The rebound effect shows up everywhere we try to introduce new technology to tame consumption. Increased energy efficiency in refrigeration, air conditioning, and internal combustion engines, for instance, seems only to have fed demand for bigger cars and fridges, and more widespread indoor cooling at lower temperatures.

Cheap delivery unleashes a similar cycle of falling prices and spiraling consumption. Once again, the pizza business offers a valuable case in point. When Domino's put the technology of telephones, television advertising, and computer maps to work, it slashed the cost of selling pizza and schlepping it to customers' homes. Prices fell, to well under $10 a pie. And we were soon eating more pizza than ever—per capita consumption of mozzarella cheese in the US leaped by a factor of seven between 1970 and 1994.

This much Jevons predicted, in the first part of his analysis. But what's happening in those Meadowlands warehouses isn't serving merely to expand existing consumption. This is *new* consumption. Rented dresses, remote storage, and half-made meals—these services have never been available to most people. My wife used to be happy wearing the same dress to every cocktail party. Now, Rent the Runway means she has to have a new one every time. I used to be content with a simple sandwich or leftovers most nights. Now, Blue Apron has me sourcing the supply chain three times a week for dinner. And it's all setting a dangerous precedent. I only wonder what's to come when AVs beat a path to my front door. More clothes, more food, more kipple ... all of it piling up in more cardboard boxes.

This is not a sustainable path. Sure, the web of continuous delivery has its green lining. Every garment borrowed for a day from Rent the Runway is one less that must be handmade only to spend its future mostly hung in a closet. Every box in a MakeSpace warehouse is one that lets us cram more microapartments into high-efficiency Manhattan high-rises. And every premeasured meal purveyed by Blue Apron produces zero food waste in my kitchen. Most important of all, these deliveries stop me from taking a trip by car several times a month—and even after accounting for the diesel-spewing trucks that bring this stuff home for me, we all end up way ahead on the carbon counter.

I tell myself this self-delusional story often, though I stop short of wrapping this remote-controlled retail extravaganza in labels like *the sharing economy*. It's hard to imagine anything less communitarian than renting haute couture, outsourcing storage for a household hoard, or breaking down bulk foodstuffs into millions of little plastic pouches. And then the rebound effect enters the analysis, and blows away all these incremental gains in ecological efficiency. For every step toward conservation that continuous delivery takes us forward, we'll take two steps back in added consumption, it seems.

If we're to tame the juggernaut of automated delivery, a big source of inspiration is the growing campaign behind *circular economies*. More a design movement than a solid theory, circular-economy thinking argues for replacing single-use, extractive methods of production with multiple-use, regenerative methods. For example, every time you compost your food waste to fertilize your garden, as I do, you're creating a tiny circular economy at home. Waste from one process feeds another, and you close the loop, conserving raw materials and energy.

Circular economies are already everywhere in our communities, but often they don't show up on the books. Because these types of exchanges exist in a kind of shadow trade outside traditional markets, they are often ignored or undervalued by economists. And their potential goes way beyond backyards. Engineered with care, and expanded to entire industries, circular economies could unleash their own type of rebound effects—instead of accelerating the depletion of natural resources, the side effect of our everyday commerce would replenish them instead.

The ultracheap, ultrafast local distribution promised by AVs offers exhilarating possibilities to give the transition from extractive economies to circular ones a boost. What if craft producers, urban farms, and other regenerative local enterprises could move material as easily as Amazon and Alibaba do? Could they turn continuous delivery and the rebound effect of cheap shipping into a force for good?

Food delivery will be an important starting point. Analysts at McKinsey predict that by mid-decade, AVs will make more than 300,000 instant deliveries in every major German city, every day—some 100 million meals annually. Today, in China, Alibaba operates dozens of massive grocery stores, and has plans to launch up to 2,000 more by 2023. Inside, "made-to-order meals gathered by store attendants from shelves and nearby cooking stations" are "wafted on aerial conveyor belts into a storeroom ... packaged and whisked to Shanghai homes within a 3km radius, at any hour and in under 30 minutes." It's a spectacular finish to the meal's million-mile, carbon-spewing journey through a thoroughly automated supply chain spanning the globe. In Montreal, however, a more sustainable model is moving forward. There, Lufa Farms provides personalized, online ordering and next-day delivery of fresh produce sourced year-round from local producers and a 63,000-square-foot rooftop greenhouse. It is a regenerative alternative that, with AVs providing cheap last-mile delivery, is entirely replicable around the world.

Why stop there? We could build on innovations like Restaurant Day, which bills itself as a "food carnival," to achieve a more circular kind of commerce. Held each year in late May, Restaurant Day encourages people in cities around the world to set up impromptu cafés out of their homes, to meet neighbors and enjoy cooking and dining together. It's easy to imagine a community-run instant-delivery service that would allow any would-be chef to run a ghost kitchen from home. Urban farms likewise would find many uses for AVs to intensify production, increase crop yields, and reduce costs of moving goods to market. Compost conveyors could become a commonplace sight, recycling organics within neighborhoods to keep carbon overhead low and food-web efficiency high.

Clothing is another huge opportunity for AV-powered circular economies. The manufacturing of clothing accounts for as much as 8 percent of global carbon emissions. What's more, textiles make up 5 percent of the waste New York City sends to landfills every year, and recycling this material is prohibitively expensive. But, as we've seen, people are eager to experiment with borrowing clothes rather than buying them. What if cheap shipping lowered the bar beyond luxury garments, and everyday clothing could make the rounds as well? A brisk trade in rented clothing might blossom as people exploit the opportunity to downsize yet constantly curate their wardrobe. Japan's AirCloset is already trying to make it work. Subscribers receive three hand-picked outfits at a time, and can swap them out whenever they want a wardrobe refresh. Since launching in 2015, the company has grown to serve some 120,000 customers in the Tokyo-Yokohama region. Imagine a future world where a great deal of the freight moving around neighborhoods on any given day is apparel and accessories. It's a perfect task to keep the robots on the night crew busy while we sleep.

Circular economies for sharing clothing could support a return to more local production of garments, too, as local manufacturers capitalized on a dirt-cheap community freight grid. Imagine grassroots industrial districts where sidewalk conveyors link up a virtual assembly line. With AVs shuttling partly-finished pieces around, small workshops and studios could be stitched together into larger supply chains, achieving a factory-like scale with a collaborative cluster of independent makers. They might even tap local vertical farms for the natural fibers needed to weave the textiles and fabrics this cottage industry consumed.

Local, circular economies won't simply be an ethical alternative to well-capitalized e-giants like Amazon. With access to the same cheap, reliable, local distribution that AVs will make possible, they could out-innovate the big guys, too, by bringing back old-fashioned hands-on service, albeit with a high-tech twist. Today, 25 to 40 percent of the apparel ordered online is returned. But it's easy to imagine a cadre of clever cyber-seamstresses cornering the neighborhood market by selling bespoke garments at factory-beating prices, sourcing materials and piecework with the help of twenty-first century logistics. Not only would they cut down waste, they'd supplant the high-performance wealth extraction of continuous delivery with a self-sustaining network of local, women-owned enterprises instead.

The End of Work?

The stars of *éX-Driver*, a Japanese anime series that first aired some 20 years ago, have a special set of skills. In a future Tokyo, long after the driverless revolution, Lorna, Lisa, and Sōichi are the only ones left who still know how to drive. The young gang's unique training comes in handy as they spend their nights hunting down rogue AVs that have succumbed to sabotage or just bad code. Like most works of the same genre, the dark underbelly of the future is glossed over with teenage themes and cartoon expressions.

Fast-forward to today, however, and alt-culture images of the ghost road are far more sinister. In *Neo Cab*, a video game released in 2019, you play the part of Lina, a taxi driver, one among the select few that remain in a city overtaken by automation. Your job: scratch out a living as a hack while searching the city of Los Ojos for your lost friend and now fugitive, Savy. "It doesn't feel like a clichéd or dramatized version of the future," one reviewer wrote in the tech blog *Gizmodo*. "If anything, it feels more like a warning sign drenched in neon."

We've feared the labor-destroying potential of machines since the earliest days of industrialization. But as automated mobility takes hold, professional drivers will take a leading role in these dystopian visions.

OUR PRESENT ANXIETY over automation's impact on jobs has a much more rigorous foundation than past panics do. Even as *éX-Driver*'s artists were drafting their dystopian visions, a pioneering collaboration of scholars at MIT and Harvard were sketching out a methodical way of thinking about the "computerization" of work (as it was then quaintly called).

Back in the late 1990s, decades of technological change had accumulated in the economy. People knew that work was changing from the widespread use of automation, but there was no theory to guide researchers or even a common language with which to frame questions. As concern among policymakers grew, an interdisciplinary dream team came together to tackle the issue—David Autor, a professor of business at MIT's Sloan School of Management; Frank Levy, a labor economist in the urban planning department of MIT's architecture school; and Richard Murnane, a Harvard-based expert on the economics of education. Over the course of the next few years, the trio worked out an ingenious and comprehensive new approach that they called a "task model."

The task model drew two fundamental distinctions. The first was between *routine* and *nonroutine* work. Routine work consists of tasks that repeat the same unchanging procedure; therefore, "they can be exhaustively specified with programmed instructions and performed by machines," the professors explained. Nonroutine work, in contrast, involves problem-solving and complex communications. "There are very few computer-based technologies that can draw inferences from models, solve novel problems," they wrote, "or form persuasive arguments." The second distinction was between tasks that primarily involve manipulation of the physical world—*manual* tasks—and those that operate in more abstract analytic and interactive realms, or *cognitive* tasks.

Armed with these two lenses, it was possible to classify how repetitive and intellectually demanding any particular job might be. Used together, the lenses isolated four distinctly different types of work—each of which, the professors reasoned, had radically different potential for computerization (see Table 6-1). Routine work of any kind was the most at risk of automation—whether it involved tasks like loading boxes onto a delivery truck (routine, manual) or filing legal records (routine, cognitive) But driving that same truck along a busy street (nonroutine, manual) and writing one of those legal documents (nonroutine, cognitive) were tasks that couldn't easily be computerized, or so it seemed at the time.

The task model provided a much-needed structure for subsequent studies and debate. For the first time, we had a vocabulary to cut through the anecdotes, hopes and fears, and binary thinking around automation's impact on work. In place of these unfounded assumptions, the model provided at least four distinct logical possibilities, including one that predicted strong synergies for humans and computers working *together* on the most challenging kinds of tasks. While the rigor was welcome, the task model was also an early warning of how complex and uncertain automation's impacts would be.

Table 6-1. Predictions of the task model for computerization of work

		ROUTINE TASKS	NONROUTINE TASKS
ANALYTIC AND INTERACTIVE TASKS			
	Examples	record keeping calculation repetitive customer service (e.g., bank teller)	forming/testing hypotheses medical diagnosis legal writing persuading/selling managing others
	Computer impact	substantial substitution	strong complementarities
MANUAL TASKS			
	Examples	picking or sorting repetitive assembly	janitorial services truck driving
	Computer impact	substantial substitution	limited opportunities for substitution or complementarity

Source: David H. Autor, Frank Levy, and Richard J. Murnane, "The Skill Content of Recent Technological Change: An Empirical Exploration," *Quarterly Journal of Economics* 118, no. 4 (November 2003), 1279–333.

The task model's conceptual framework has shaped how today's researchers make sense of automation's prospects for human work. But its specific predictions have proved remarkably prescient and surprisingly durable as well. When the task model was first published in 2003, much of the routine, cognitive tasks that Autor, Levy, and Murnane identified were already heavily automated. ATMs had decimated the ranks of bank tellers, and databases had destroyed clerking. But the authors' predictions for the automation of "picking or sorting" (routine, manual) are only now playing out in e-commerce fulfillment where, as we saw earlier, warehouse droids have replaced human workers. Meanwhile, the more optimistic forecasts about complementarities have shown up, too. The use of computer vision in medical diagnostics is helping doctors deliver better care at lower cost. And the business of persuasion is one where human communication has been extensively amplified by AI, for better or worse. One thing the researchers couldn't anticipate, however, was deep learning. The technology's rapid advance shifted the technology trend line. And while the task model tags truck driving for its "limited opportunities for substitution or complementarity," investors disagree, pouring an estimated $1 billion into self-driving truck startups in 2017 alone. Even if you don't buy the likelihood of fully automated trucks, in keeping with the task model's projections, computerized copilots are definitely on the way.

Despite its missed predictions, for a first crack at a truly wicked problem, the task model was a big step forward. (The authors, however, described their work as merely an "informal" attempt to chart the territory.) It would take more than a decade before the model could truly be put to the test in a comprehensive, detailed teardown of the entire labor force, job by job. And during that period advances in machine learning pushed computerization forward not only in industry but in the scientific community as well. Ironically, it was the arrival of cheap AI in the economics lab that would reveal how automation threatened a far wider range of jobs previously thought safe.

The nonroutine, cognitive work of economics research turned out to have strong complementarities indeed, much to the benefit of Carl Benedikt Frey of Oxford University. A dozen years after the task-model paper was first published, Frey's group harnessed machine learning to vastly improve the depth and breadth of its predictions. Their hybrid work of human-machine collaboration involved two steps. First, working with AI experts from Oxford's engineering school, Frey and his team combed through a mass of job descriptions in O*NET, a database detailing the various tasks involved in more than 700 types of jobs. Working with a sample of 70 occupations, the panel assigned each task a subjective "automatability" score ranging from 0 (not automatable) to 1 (fully automatable). This "training set," as it's called in machine learning, was then used to calibrate a model that could make educated guesses about the automatability of the other 600-plus vocations in O*NET.

The results of this AI-fueled speculation were stunning. According to the Oxford group's analysis, some 47 percent of jobs in the US economy were at risk of automation over the next two decades. The report triggered a wave of sensational reporting, and captured the attention of workers and policymakers the world over.

For all the hit-you-over-the-head headlines, the most revealing insights about the driverless revolution's impacts on work were buried in the details of the Oxford study. To begin with, it puts the MIT-Harvard gang's miscalculation on truck driving in perspective. As it turns out, there's more than one kind of truck driver. O*NET's occupational classification system identifies no fewer than three different types of truck drivers: industrial truck and tractor operators, heavy and tractor-trailer truck drivers, and light-truck or delivery-services drivers.

What kind of truck driver you are makes all the difference for your future career prospects. The results map almost perfectly to how we've seen automation progress in trucking—from factory and field, to highway, to city street. As with all the 702 jobs in the study, each truck-driving occupation was ranked in order of proclivity for automation (rank 1 being the least computerizable, rank 702 the most) based on a numerical score ranging from 0 to 1. Industrial truck and tractor operators come in high (rank 588), consistent with the widespread automation of mining, farming, port, and warehouse vehicles we've seen to date. This job class has a 0.93 probability of computerization, according to Frey and company, making it an all-but-certain outcome. Heavy and tractor-trailer truck drivers (rank 431, probability 0.79) are still highly vulnerable, but as we've seen, there's ample work for future human crews on the highway of heavy hauling.

It's in the last mile, though, fraught with nonroutine challenges, where truck drivers will hang on the longest. Light-truck or delivery-services drivers (at rank 380, probability 0.69) are only somewhat more likely than the average worker to fall victim to automation.

THE TASK MODEL tells us an awful lot about how automation could impact the work we do in today's economy. But by focusing solely on existing occupations, it boxes us into a position where all we can see of the future are the remnants of today's workforce. Read the Oxford study, and you'll walk away convinced that the future doesn't look good for humans.

But what if the jobs of tomorrow are not the jobs of today? That's one of the questions asked by another blue-ribbon study, *America's Workforce and the Self-Driving Future*, commissioned by Securing America's Future Energy (SAFE), a nonpartisan group that seeks to reduce US dependence on imported oil. SAFE has been a strong advocate of AVs as a technology to reduce fossil fuel consumption and achieve the group's goal more quickly. Published in 2018 and looking out as far as 2050, SAFE's report zeroes in on the employment impacts of self-driving technology. It is as optimistic as Frey's work is dismal, describing a country where the impacts of automated vehicles on employment are all but "unnoticeable" until well into the 2030s and even then are manageable, not catastrophic. And while the Oxford economists see professional drivers in the crosshairs of automation's advance, SAFE's experts argue that the structural increase in unemployment—even at the peak of impact, around 2040—will be less than 0.2 percent. Instead of massive, irreplaceable job loss, they forecast AVs will deliver more than $700 billion in social and economic benefits annually by 2050—through congestion mitigation, accident reduction, lower oil consumption, and the recovered value of time spent driving. According to SAFE, any one of these benefits vastly outweighs projected wage losses, estimated to peak at $18 billion in 2044 and 2045, by a factor of three or more. The savings, it is presumed, will be reinvested broadly throughout the economy, spurring the creation of jobs to replace those lost to AVs.

SAFE's dream of an AV-powered golden age is as energizing as the Oxford group's nightmare of the end of work is terrifying. (Notably, MIT's David Autor, one of the original creators of the task model, endorsed the SAFE findings.) It says that automated freight and continuous delivery are an economic engine waiting to be revved, rather than a juggernaut coming to crush what's left of the working middle class. What's more, there's still plenty of time for policymakers to prepare the workforce and businesses to make a smooth shift. Whether SAFE's optimistic forecast is seen as a call to action or an excuse for complacency will be a political decision. But compared to the Oxford doomsayers, it seems more action-able. The report offers three possible approaches for the US, before settling on a recommended policy that would channel some of the economic gains from AVs directly into workforce-development programs.

Until automation shows its job-sapping hand, this measured approach makes sense. Advances in freight transportation have always wreaked havoc on cities and neighborhoods. The change is often painful. Wharves in Brooklyn and San Francisco gave way to container ports in Newark and Oakland, for instance. Tens of thousands of jobs disappeared every year, for decade after decade, as these cities deindustrialized. But over time, the cities recovered and thrived—employment, earnings, and wealth all exceeded their previous highs. No one laments the lost longshoremen of the West Side anymore except when they watch *On the Waterfront* (which was mostly filmed across the river in New Jersey, anyway). They picnic on those piers and surf the web with free Wi-Fi instead.

Rethinking Risk

There's one more twist in this tale of no-man's-land, because ruthless efficiency isn't the only game in town. Sometimes setting aside something for a rainy day is the best strategy. While cutting costs will drive much of the business case for investing in AVs, automated shipping will also change how companies cope with the growing uncertainties of the twenty-first century. Near-zero-cost freight will make organizations of all kinds more resilient to many types of business risks—even as those risks increase.

Take road safety—the single greatest source of operational risk for many companies that rely heavily on transportation. Today, about 4,000 Americans are killed, and 100,000 injured, annually by trucks and buses. In the coming years this number is certain to increase as the trucking workforce ages and distracted driving rises. The numbers are staggering. Over the next decade, the American Trucking Associations expects 400,000 seasoned drivers to retire and as many as 900,000 younger recruits to enter the trucking workforce. Yet truck drivers age 21 to 24 are involved in accidents at twice the rate of those age 30 and older. Insurance premiums are already soaring in anticipation of these growing risks, and are likely to rise much higher. By eliminating or reducing the severity of crashes, AVs won't merely save lives—they'll bring medical and property damage claims under control too. Additionally, copious streams of data collected by commercial AVs will power pay-per-mile insurance services such as Rideshur, a London-based startup whose software models accident risk exposure down to a 300-meter resolution and adjusts rates every 200 milliseconds.

Automated fleets also offer new opportunities for firms to become more agile in managing economic risks. Faster deliveries, made by tireless robodrivers, mean less exposure to delays and spoilage caused by hazardous weather and traffic. Test drives of a modified tractor-trailer by Embark, a self-driving startup, demonstrated that fully automated AVs could cut the cross-country travel time from California to Florida from five days to just two.

Energy costs are a particularly volatile threat. Diesel fuel can account for as much as one-third of trucking companies' cost of doing business, and even a small shift in price or efficiency can make or break the bottom line. Again, the changing labor force will create more liability for companies. Young drivers are more inclined to rapid acceleration and deceleration, and use 30 percent more fuel than experienced hands. But with a computer at the throttle, every truck could achieve the fuel efficiency of expert drivers. What's more, as climate change disrupts energy infrastructure in the coming decades, AVs will be better equipped to route themselves around future infrastructure failures. Much as the internet heals itself, directing data around broken links, we can imagine truck fleets that automatically adjust routes to avoid regions that are short on gas, sun, or whatever makes them go in the future.

Unsold inventory is another existential risk where automated mobility will provide companies with new options. For decades, just-in-time approaches have sought to reduce stockpiles of goods to the bare minimum needed for the next cycle of production or distribution. AVs will play a major role in the next wave of slimming down stocks. Walmart, for instance, has already trimmed billions of dollars of inventory using automated systems for loading and unloading delivery vehicles and restocking shelves. But a more radical approach to automated inventory management could do away with warehouses altogether. Imagine a fleet of AV freighters synchronized like a microtransit mesh. The inventory that used to be stored in a central depot would instead be spread out among various vehicles. To get a needed item from its current rolling location to your front door would entail a series of transfers, like a passenger riding driverless shuttles across town. Goods would flow like hot potatoes from AV to AV to their final destination without stopping over in stationary storage.

As an added benefit, a warehouseless distribution system would eliminate the risks created by having a fulfillment center as a single point of failure. In February 2019, as executives of UK online grocery chain Ocado reported quarterly earnings figures to a roomful of investors in London, the company's automated warehouse in Andover was burning to the ground, its 600-strong android workforce trapped inside. It took more than 200 firefighters, who were forced to cut holes in the structure's roof to fight the blaze, some three days to fully extinguish the flames.

The widespread use of robofreight will, however, create new systemic risks. In the last decade we've witnessed how automated trading in financial markets can flood exchanges with electronic orders and produce flash crashes. Imagine that kind of power put to work in the world of distribution. A speculator might reach out from across the globe and drop-ship a factory's worth of goods onto city streets simply to undercut the competition. We'd have to revise the dictionary entry for *dumping* to reflect the arrival of this barbaric new trade practice. Or could companies withdraw from communities at a moment's notice if their demands for tax breaks, infrastructure, and regulatory concessions weren't met? The arbitrary creation of such sudden abundance or scarcity, simply to underwrite speculation, would be at best unseemly, at worst immoral. But if this weapon of mass commercial destruction were put to use, powered by dirt-cheap automation, its payload would arrive with terrifying stealth and speed.

PART III

Taming the Autonomous Vehicle

7   The New Highwaymen

The purpose of a system is what it does.

—Stafford Beer, father of cybernetics

The road from Derby to Chesterfield in England's East Midlands was once the world's most dangerous. As the English Civil War drew to a close following the execution of Charles I in 1649, Royalist officers who were demobilized during the denouement suddenly found themselves with no means of support. A flamboyant few, like Dick Turpin and Swift Nick, took to the isolated hills of the Peak District, terrorizing and robbing travelers on the valley roads below. The "highwaymen," as they were known, were feared across the land. While these brigands live on in legend, the last mounted heist in England took place in 1831. Many highwaymen had become the target of gruesome reprisals by the locals. One popular punishment was gibbeting, which involved stringing up the villain's mutilated body for public display at a crossroads. As it turns out, the highwaymen weren't the only ones looking to cash in on the travel boom that swept Britain in the early years of the Industrial Revolution. A legal racket was moving in. As the nineteenth century sped on, the volume of intercity travel grew quickly. Heavier carts and carriages, moving in ever-increasing numbers, clobbered old paths built over the centuries by local parishes. To coordinate construction and maintenance of more robust roads, local "turnpike trusts" were established by Parliament. At their height, in the 1830s, after a wave of construction, over 1,000 turnpike trusts ran some 30,000 miles of toll roads throughout Great Britain. These locally managed organizations issued bonds and levied tolls to finance them—in 1838 taking in some £1.5 million against more than £7 million in debt. Seemingly as benevolent as the highwaymen were predatory, the turnpike trusts boosted property values and helped local economies grow through trade.

The turnpike trusts, however, introduced a new kind of institutionalized highway robbery. More than 8,000 toll gates and side bars—smaller gates blocking local byroads—demanded micropayment for every leg of a journey. The side bars prevented travelers from avoiding toll gates on the main road, but they angered farmers, who often depended on the blocked shortcuts to access their fields.

In Wales, where a deep economic depression had taken hold, the turnpike trusts became the target of widespread civil unrest. In the 1830s, even as farming families struggled to make ends meet, wealthy landowners continued to raise rates for passage on local roads. Adding to the injustice, "many of these landowners served as local magistrates and on 'Poor Law' committees," which administered the system of public relief put in place by Parliament in 1834 and immortalized in Charles Dickens's novels. This "meant they were basically able to determine and enforce their own toll charges, without contention from government."

Tensions came to a head in 1839, when Thomas Rees (Twm Carnabwth in Welsh) led an angry mob of men disguised in women's clothing to the toll house at Yr Efail Wen, in Carmarthenshire, and torched it. As the despised gate burned to the ground, the men shouted a verse from the Book of Genesis, "And they blessed Rebekah and said unto her, Thou art our sister, be thou the mother of thousands of millions, and let thy seed possess the gate of those which hate them." This highly symbolic gate, however, was quickly rebuilt by the local turnpike trust, prompting the gang of cross-dressers to sally forth and torch it again a few months later.

The "Rebecca Riots" continued for four more years, growing to involve thousands of men—most of whom continued to revolt in drag. But despite their peculiar style of protest, and the fact that the toll houses were but the most visible symbol of greater grievances—the Rebeccas' uprising is a powerful reminder of a universal truth. Once roads are furnished gratis, any later attempt to impose fees will provoke a spirited reaction.

It was a lesson well learned by the road builders that followed. The turnpike trusts faded away, done in by the rise of railroads rather than the Rebeccas. With the turnpikes' parochial barriers to the free flow of people and goods removed, their operations and debts were consolidated over larger districts. The private trusts were replaced by public road authorities, and by the early twentieth century the true costs of roads came to be financed through more indirect methods, such as fuel taxes. To this day, to most drivers in most places on earth, roads are understood for all intents and purposes to be *free*.

Free roads, however, will soon become an endangered species— yet another victim of mass automation. When every movement can be tracked, we'll pay per mile and fraction thereof. The turnpikes' preference for micropayments now seems prescient, indeed. But we won't be going back to the days of cozy community trusts. With every vehicle directly connected to the cloud, local money will be replaced by global capital, policy by markets, and public ownership by private control.

This brings us to the third big story of the driverless revolution, a process I call the *financialization of mobility*.

One of the more confounding outcomes of globalization is that even as the world's workers have seen wages stagnate, the share of surplus wealth claimed by bankers has swelled. In the US, in 2010 the total compensation of financial intermediaries reached an all-time high of 9 percent of GDP—some $1.4 trillion, up from less than 5 percent before

1980. And that was *after* the financial crisis blew up a big part of their paper profits.

The rapid expansion of moneyhandlers' size and influence, a process that critics call *financialization*, has ensnared critical sectors of the material economy that for much of the twentieth century were sheltered by custom or regulation from full market pressure. Residential mortgage debt, largely unknown in America before World War II, ballooned from 15 percent of GDP in 1948 to more than 80 percent in 2018. As much as half the price of a barrel of oil is attributable to speculative trading. And in 2008, farmers produced enough food to feed the world's population twice over, yet more people starved to death that year than ever before—victims of a systematic effort by commodities traders to manipulate markets for staples like wheat, corn, and rice. Most of the dead were farmers.

Transportation's turn at the carving table of high finance is long overdue. While it may begin in benign fashion—as governments exploit the driverless revolution to slip in new tolls, taxes, and fees that claw back the true cost of free roads—it will quickly take a malevolent turn. Even as automation helps unwind the bad policies of the motor age, it will open the door for an insidious infiltration of markets into every choice about when, where, and how we travel. This new logic will profoundly shape our behavior, and not always for the greater good.

This marks a critical turning point in our journey on the ghost road. Up until now, automation has mostly been working for us, delivering an abundance of new products, capabilities, and services (Part I). Or it was working around us, reshuffling our material world to deliver a sustainable standard of living our ancestors never even dreamed of (Part II). But in these final chapters, we'll consider a much darker possibility. What happens when automation works directly against us?

It's here, as we confront the financialization of mobility in the driverless revolution, that the stakes run highest. Will we allow markets to supplant democratic deliberation in deciding who gets to travel when, where, why, and how? And how do we ensure affordable, reliable, sustainable transportation when the tail of global finance wags the dog of urban mobility?

The answers to these questions will mean the difference between who gets ahead in the AV-powered future and who's left behind.

The Metered Mile

If the age of free roads is coming to an end, the idea that may finally spell its doom is a scheme known as congestion pricing. By charging people more to travel when and where traffic is heaviest, the thinking goes, people will switch to mass transit, shift their schedules, or take different routes. It's the darling of think tanks, an idea opposed only by self-serving motorists. But it will also be a catalyst for mobility's financialization.

As legend has it, Columbia University economist William Vickrey, the father of congestion pricing, was so "uninterested in material comfort that he barely knew how much he was paid." The absentminded professor was well aware of money's motivating power, however—auctions were his primary scholarly interest. Vickrey's pioneering work on the subject earned him a Nobel Prize in 1996. But the polymath was also keenly interested in using market economics to improve the way we allocate public resources. Early in his career, New York City would present him with an ideal opportunity to test his ideas.

It was a fitting, yet ironic, match of man, market, and metropolis. Capitalism had created the great industrial city and, by 1951, was in the process of hollowing it out. The subway system was in rapid decline as the city's population steadily drained into newly built suburbs. Ridership and revenue had peaked, and the transit authority's financial outlook was grim. The previous year, a police-corruption scandal had forced the resignation of Mayor William O'Dwyer. In an effort to restore public trust, his successor, Vincent Impellitteri, ordered a sweeping review of the city's finances. Vickrey was chosen to lead the assessment of transit fares and to try to find a way to stabilize the system.

The rambling study Vickrey produced for City Hall spans more than 150 pages, and makes for dry reading. But it reveals the conceptual roots of congestion pricing as it is practiced today. Vickrey understood the problem foremost as an exercise in marginal-cost pricing, where the price of goods and services is set at the additional cost involved in providing them, rather than at what the market will bear. Fixed costs like track and trains are ignored. In business, marginal-cost pricing often serves as the basis for deep discounting during periods of slow sales—allowing a producer to set the lowest price possible while maintaining output and avoiding operating losses. For the New York City public-transit system, Vickrey reasoned, marginal cost was a perfect measure of what riders would be willing to pay to avoid the delays and discomfort of congestion.

Following this insight, the bulk of the 1951 report documents Vickrey's painstaking calculation of the spending necessary to relieve rush-hour jams—costs such as adding trains and expanding stations, which grow at an increasing rate as crowding gets worse and worse. The final tally, a detailed matrix of the marginal costs for travel between various zones at different times, was stunning. Vickrey demonstrated that to fairly reflect the marginal costs of rush-hour service into Manhattan's central business district, fares should be 25 cents—five times higher than the prevailing flat fare of the day (Table 7-1).

Revolutionary in thinking and epic in scope, Vickrey's pricing proposal utterly perplexed New York City's politicians. They balked at the complexity of fare collection the scheme would require. But Vickrey had a well-considered plan for this detail, too. Upon entering the system, riders would pay a deposit of one quarter (25 cents), and fares would be capped at 20 cents. The remaining five cents was held as a refundable deposit, to be collected as riders exited the system, minus a fare based on the time and zones of travel. More than 20 pages of text and drawings detailed not one but two systems—one purely mechanical, another electronic—for the accurate collection of time- and zone-based fares. But despite his system's promise to reduce overcrowding and increase revenues, Vickrey's ideas

Table 7-1. Estimated marginal cost of New York City subway traffic, 1952

	MARGINAL COST OF TRAVEL TO CENTRAL ZONE (IN CENTS)	
	A.M. RUSH	NONRUSH
FROM		
Outer zone	25 cents	5 cents
Middle zone	20 cents	5 cents
Central zone	10 cents	5 cents

Source: Adapted from William S. Vickrey, *The Revision of the Rapid Transit Fare Structure of the City of New York*, Technical Monograph No. 3, 1952.

were simply too far ahead of their time. In his lifetime Vickrey would see only scattered application of marginal-cost pricing to subways. As one biographer notes, "Transit pricing still appears to be set according to the political and budgetary pressures of the moment."

Congestion pricing fared somewhat better on surface roads. In 1959, Vickrey was invited by the US Congress to prepare a road-pricing scheme for the Washington, DC, region. As ordered, the professor submitted another characteristically thorough report, but once again the plan was not pursued due to perceived political and technical infeasibility. When Vickrey died of a heart attack in 1996—behind the wheel of his own car on the Hutchinson River Parkway, no less—the world had but one successful example of congestion pricing to point to, Singapore's Area Licensing Scheme. Singapore's program had, however, delivered on Vickrey's promises. After it launched in 1975, traffic into the island nation's central cordon zone during tolled hours fell by 44 percent, more than the 30 percent projected. And as the share of commuters traveling by private car fell, those using carpools and buses expanded from 41 to 62 percent.

From Singapore, congestion pricing has spread slowly but surely. London (in 2003) and Stockholm (in 2006) were the next two big cities to put the scheme to work, and after introducing congestion-pricing cordons around their central districts, both saw traffic fall by nearly one-third. In all three cities, authorities have cut back on road construction and maintenance and earmarked toll revenues for public-transit improvements. The air is cleaner, producing measurable public-health benefits that even Vickrey didn't anticipate—among the biggest beneficiaries are millions of children who'd otherwise be at elevated risks of emissions-induced asthma. A fourth city, Milan, adopted congestion pricing in 2012, and New York City began preparations to introduce it in 2019.

Despite these successes, the spread of congestion pricing is just beginning. In many more places, motorists have resisted these reforms through organized lobbies. They've rolled out upside-down arguments about equity, claiming that congestion tolls would hit the middle class and poor the hardest. But in fact, it is the poor who benefit the most from congestion pricing. Not only are they the least likely to own cars, they gain the most from faster buses running on traffic-free streets.

The uptick in congestion pricing's appeal, however, reflects a tipping point for traffic jams. It turns out that the biggest obstacle to broader support for congestion pricing was insufficient congestion. Over the last decade, the surge of ride-hail vehicles into center cities has brought a new intensity to the traffic crisis. Public opinion is shifting fast. Take San Francisco, for instance, ride-hail's birthplace and the city where people have most eagerly embraced it. By 2016, one-quarter of all vehicle congestion citywide was blamed on Uber and Lyft's fleets. Fully one-half of the increase in traffic since 2010 was attributed to the ride-hail giants. More alarming than the overall trend were the localized spikes. In the city's downtown financial district, for instance, Uber and Lyft accounted for a whopping 73 percent of the increased traffic in recent years. Reprogramming mobility had eliminated a decade's worth of painstaking work by transit agencies, cycling advocates, and walkability planners to reduce auto use. It was a warning of what was to come for other cities as well.

What's more, the political insurmountability of congestion pricing doesn't seem so insurmountable anymore. London's efforts showed that the approach wasn't only viable for centralized states like Singapore or social democracies like Sweden. You could do it in a big, messy, global city and it could work. And even in the perennially dysfunctional politics of Italy (and New York City, for that matter), a consensus for congestion pricing could be found.

WE AREN'T THE FIRST to tackle the challenge of tracking the passage of vehicles. Taxi meters, which log the distance traveled in order to compute a proportional fare, are as old as Western civilization. Greek mathematician Archimedes first imagined such a device in the third century BC. A few centuries later the great Roman architect Vitruvius finally built it. His box of gears—linked at one end to a chariot's axle and at the other to a small dial—regulated the sprawling empire's ride-hail network, the *cisia*. Vickrey obsessed over the problem of toll collection, too, proposing the use of wireless tags to automate toll collection (he hated the idea of *adding* congestion through the construction of toll gates). When critics deemed the scheme impracticable, the tireless professor constructed a prototype on his own, giving himself a parts budget of only three dollars for each device.

Such contraptions are no longer necessary, because *we* have become the tracking devices that meter the movement of our machines. As we move through the world, our smartphones leave behind a trail of digital breadcrumbs that precisely pinpoint the position, velocity, and type of vehicle we're in. Instead, engineers today are focused on finding new and ever-more-advanced ways to tally the bill. It's no longer a simple function of distance times rate.

One startup that's at the forefront is ClearRoad, a Brooklyn-based maker of road-pricing software. ClearRoad doesn't make AVs, put sensors on roads, or operate any transportation infrastructure of its own. Instead it runs the network that connects governments that own roads, companies and individuals that own vehicles, and banks that hold everyone's cash—all for the purpose of high-precision tolling. Think of it like an internet router, but for miles and money.

Let's say one morning in the not-so-distant future your car is driving you to work on the freshly renamed Santa Monica Tollway, which links the seaside city with downtown Los Angeles and points east. Since you've signed up for electronic tolling, your position is being logged second by second, and you've given permission for that data to be sent to ClearRoad, which calculates your fee based on how many miles you travel. Funds are automatically drawn down from your linked bank account, and your paltry payment is bundled with thousands of other drivers'—less a small service fee that ClearRoad skims off for its trouble—and paid out electronically to Caltrans. The appeal of this arrangement is its flexibility. ClearRoad handles all the difficult work of tying the pieces together, letting everyone use the vehicles, hardware, software, and banks that suit us best. Government doesn't have to get involved with you, and you don't have to get involved with the government. You won't even need to use its toll tag. Your car, or your phone, or both, can track the trip instead.

Part virtual tollbooth, part clearinghouse, ClearRoad's platform makes it painless to charge drivers for every mile they drive on roads that used to be free. In the short term, this provides an escape route for governments to wean their road authorities off fuel taxes, which are bound to disappear with the rise of electric vehicles. In the long term, ClearRoad's approach provides a flexible framework for tracking and tolling any machine the driverless revolution dreams up. It is policy agnostic, equally able to impose either congestion-based fees or solely distance-based ones.

"Basically, it's a ledger," explained Frederic Charlier, ClearRoad's founder and CEO, during a 2018 visit to the team's office at Urban-X, a high-tech incubator in Brooklyn's hip Greenpoint neighborhood. "On one side is the mileage; the other side is money—accurate, indelible, and secure."

The importance of what ClearRoad is building is undeniable. Putting aside the politics of congestion pricing, there aren't many options for governments to replace fuel taxes. This at least eases the pain for drivers during the transition to use-based charges. But the business of back-end integration didn't excite me. It felt incremental, not transformational. I closed my notebook, and looked for the door.

Then Paul Salama, the company's chief evangelist, spoke up. "We're creating the financial infrastructure for road pricing," he offered.

I paused for a moment and pondered the reach of this pecuniary platform as millions of AVs swarmed the streets. Had Charlier sold his vision short by describing it as a mere *ledger*? As I opened my notebook again, he recounted his hazing in the business of road pricing, during a five-year stint at Sanef, a private company that operates the national highways in northeastern France. To recoup its costs of maintenance and operations—and make a modest, regulated profit under contract with the national government—Sanef is an aggressive and sophisticated user of electronic tolling, he explained.

As I listened, dollar signs (and euros, I suppose) danced before my eyes. I wondered if, as future highways filled up with self-driving vehicles, governments would seize the opportunity to try out more-ambitious tolling schemes. I thought about my internet-connected phone, and TV, too, and all the casually clicked subscriptions that mounted up over the months. Was that the kind of thing to expect on the roads too? I posed the possibility, and Charlier pushed back, arguing that public authorities would find it difficult to impose such complicated, ever-changing toll schedules. Politics would severely limit their freedom for financial innovation. (See the Rebecca Riots, pages 162–63.) But private concessionaires like Sanef, he confessed, could and would push the technology to its revenue-maximizing potential.

I left the meeting with an urgent lingering question—what sophisticated and possibly sinister tolling schemes lay ahead? Salama's bold assertion started to take on a larger significance. ClearRoad wasn't building the financial infrastructure for road pricing. It was building the perfect incubator for the financialization of mobility itself. On ClearRoad's servers, you wouldn't simply load some new code to change one vehicle's behavior. You could change an entire market with the press of a key. The possibilities to charge for any kind of movement, at any time, slicing and dicing the market to reap every last penny it was willing to pay, were almost endless. It could become a hothouse for financial engineering.

Governments would surely want to cash in, and the potential for perverse incentives to creep in disturbed me deeply. I imagined road managers tuning highways to find the perfect balance of traffic and cash flow, twisting knobs and flipping switches to juggle political and fiscal realities. Dial up the tolls to fund much-needed repairs, or slide them back down to soothe constituents' anger. Since AVs track every step with precision, why not charge more for congestion-causing behavior like left-hand turns across oncoming traffic? How about a futures market for prepayment of tolls? Vickrey would no doubt approve. After all, that was a key insight in his own early work, which laid the foundations for how airline tickets are still sold today. Then, parlay those financial instruments into secondary markets, and let road operators, mobility-service providers, and hedge funds engage in speculative trade against the risks and rewards of road revenues. Now we're off to the races. The road to ruin for congestion pricing—from sensible scheme for allocating public space, to funnel for financial mayhem—may be far shorter than it seems today.

Right now, an uneasy consensus exists around the efficacy of congestion pricing. For the left, this alliance has achieved two important programmatic goals—securing new funding for transit, while simultaneously discouraging automobile use. In New York City, for instance, the scheme adopted in 2019 is expected to bring in more than $1 billion a year, financing $15 billion in transit improvements. But the victory has come at a steep price in principle, because for the right, congestion pricing is a stunning ideological victory. Not only does it balance spending with user fees rather than with wealth or income taxes, it tames the public realm under the heel of a market mechanism. This harmony may yet prove naïve, as the driverless revolution pushes the wedge of financialization deeper and deeper into the world of mobility. New York City's struggles over how exactly to implement congestion pricing offer the clearest case of how fast priorities can shift from clearing roads to filling city coffers, and to fighting over which group bears the cost or reaps the rewards.

With the proliferation of more complicated and lucrative tolling schemes to tax the connected, computer-controlled cars of the future, the power struggles will only intensify. For now, most cities pipe congestion fees back into transit, to create a virtuous circle of subsidy that further reduces traffic and emissions. But already, roads are being eyed as an ATM for plugging bigger budget holes far afield of transportation. Inevitably, financial institutions will make their power felt in these struggles in the form of clever but costly schemes to capture and leverage road revenues, often by luring cities into penurious deals with up-front, lump-sum, low-ball cash payments. Congestion pricing could be the gateway drug to a future where financiers and mayors team up to tax our movements in endlessly creative and lucrative ways. If you think I'm paranoid, consider that in 2008, Chicago mayor Richard Daley signed a 75-year contract putting the city's parking meters in the hands of a consortium of sovereign wealth funds led by Morgan Stanley. "Since then, rates have skyrocketed (downtown parking more than doubled in cost to $6.50 an hour by 2013) and the company has netted $778.6 million in revenue. By 2020, the company will have made back its initial $1.15 billion investment and will continue to profit for 60 years."

When I think about congestion pricing, I *worry* about whether we're sufficiently scrutinizing the ideology it represents. But what I *fear* is whether we can contain this rationale for rationing once it has been unleashed. I don't dispute that giving away the public realm to motorists has been a terrible mistake. But as pragmatic and politically expedient as congestion pricing is today, once automation takes hold it may soon turn out to be a deal with the devil.

Return of the Traction Monopolies

The octopus—a "horrible monster, whose tentacles spread poverty, disease and death"—was the icon of choice for America's political cartoonists depicting the unprecedented reach of John D. Rockefeller's Standard Oil empire in the 1880s. Yet it wasn't until years later that this tenacious new form of anticapitalist critique reached its pinnacle. George Luks's 1899 drawing "The Menace of the Hour" portrayed the vast network of interests entangled by New York City's street railway syndicate—"the traction monster" that had held the city hostage for years. It remains one of the most searing images of the Gilded Age's dark underbelly (Figure 7-1).

When electric motors displaced horses, the new technology transformed the way Americans traveled in big cities. But the new financial requirements of electrification shook up the business of urban transportation even more fundamentally. Up until the 1880s, "public transportation in large American cities was provided by numerous, competing

Figure 7-1. *"The Menace of the Hour," 1899.*

UNIVERSAL HISTORY ARCHIVE/UNIVERSAL IMAGES GROUP VIA GETTY IMAGES.

horsecar companies. The firms were autonomous and their operations uncoordinated." In the horsecar trade, independent operators could start a small company with a modest amount of money raised from friends and family, buy a vehicle, and work for themselves. But the arrival of cable cars in San Francisco and Chicago in the 1870s, and electric streetcars in the late 1880s, dramatically increased the working capital needed to enter the business. Suddenly transportation companies had to build *infrastructure*, too.

The new environment favored not only well-financed firms but politically connected ones. Budding mass-transit entrepreneurs suddenly found themselves negotiating with corrupt municipal officials over complex contracts for access to city streets. Collusion to box out competition was widespread. Alliances with electric-power providers were also crucial to citywide control. In Philadelphia, the biggest streetcar operators and power companies merely joined forces, forming a syndicate that dominated transportation and energy. But in Seattle, they merged into one all-powerful entity, the Puget Sound Traction, Light & Power Company. Formed in 1900, it would become the most powerful, reviled traction monopoly in North America, consolidating a decade's worth of frenzied infrastructure expansion in the waterfront boomtown. In Philadelphia; Seattle; Washington, DC; Newark, New Jersey; and many other cities, the traction monopolies enjoyed decades of unrivaled power over surface transportation until a combination of technological change and federal intervention brought about their demise.

—

TODAY, URBAN TRACTION MONOPOLIES are being built once again. Instead of steam-powered generators and electric motors, they harness wireless networks and handheld supercomputers. Soon, they'll add AVs to their arsenals. But while they hold the same aspirations for market domination and follow their predecessors' familiar playbook, their scale is far greater and their reach into our lives far deeper. Their tentacles are already encircling the globe and will shape how we travel for decades to come.

Nowhere are the ambitions of tomorrow's traction tycoons clearer than in the empire assembled by SoftBank, the Japanese holding company founded by Masayoshi Son in 1981 (Figure 7-2). From its humble beginnings as a computer parts store, SoftBank has grown into the world's largest technology investor. The group's $100 billion Vision Fund, launched in 2017, wields a purse that is itself bigger than the entire venture-capital industry, which invests a mere $70 billion annually worldwide. Staking this purse on young companies that are exploiting artificial intelligence to reorganize big chunks of the physical world, SoftBank's investments span the gamut of city-building sectors—real estate, hospitality, food, and retail. But the first target for world domination—and the key to it all—is transportation.

Rarely has so much money moved so fast. The Vision Fund launched in May 2017, a few days after SoftBank announced its second infusion of cash into Chinese ride-hail giant Didi. That $5.5 billion financing brought its total commitment to some $9 billion. Then in July, Didi and SoftBank

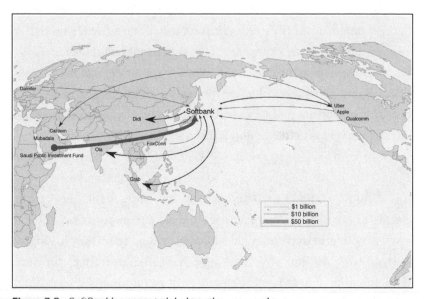

Figure 7-2. *SoftBank's nascent global traction monopoly.*

poured a combined $2 billion into Singapore-based Grab, which operates throughout Southeast Asia. October was India's turn, where local upstart Ola downloaded $2 billion of SoftBank loot to its accounts. Finally, in November, the group took a $10 billion, 15 percent stake in Uber. To top it off, in 2019 Uber picked off Dubai-based Careem, a Middle East operator, for $3.1 billion.

The octopus quickly tightened its grip. Through its web of investments, SoftBank's leadership took up seats on the boards of companies handling some 90 percent of the 45 million ride-hail trips taken worldwide every day. Immediately, it began pressing for an end to costly price wars across its extranational empire. The first and boldest move was to pull Uber out of six markets across Southeast Asia. The damage to consumers, judged by economists to be "irreversible," was substantial. With its most important rival gone, Grab cornered more than 80 percent of the Singaporean ride-hail market, and raised fares an average of nearly 15 percent. When I traveled to Singapore in November 2018, I had the eerie feeling of opening the usually teeming Uber app to find a map devoid of life. A ghost road, indeed.

Government response was swift, but inconsequential. Vietnam, Singapore, and the Philippines all launched antitrust investigations soon after the Uber-Grab shuffle. In Singapore, where the rule of law is strongest, a fine of some $9.5 million was widely viewed as a "minor bump in the company's growth plans." SoftBank took the experience as license to push on, immediately pulling Uber out of another fare-slashing battle with Ola in India. As one journalist put it, "SoftBank is playing the ride-hailing version of Risk," the classic board game of global conquest, "but it also owns a piece of all the players."

In North America, SoftBank's money financed a more costly strategy—a price war of attrition waged against Lyft, Uber's only substantial competition. In 2019, Uber and Lyft each floated initial public offerings, yet both continued to burn cash at extraordinary speed, spending more than half their current revenue on "driver incentives, passenger discounts, sales, and marketing to acquire passengers and drivers faster than the other." As Silicon Valley guru Tim O'Reilly argued, the two giants were "locked in a capital-fueled deathmatch."

SoftBank's plan was simple—have the deepest pockets for the slog ahead. The Vision Fund's biggest investor, Saudi Arabia's Public Investment Fund, was also the first to sign on. For $45 billion, the proceeds of an earlier liquidation of petrochemical giant Saudi Basic Industries, the Saudi Crown took a nearly 50 percent stake in Son's dream. Abu Dhabi's Mubadala, another sovereign investor, chipped in $15 billion more. All told, two-thirds of the cash kitty that's financing SoftBank's vision consists of the profits of a half century's exploitation of fossil fuels. Yet this may be just the beginning. Son has tapped a well of capital so deep, it is for all intents and purposes bottomless. One plan to sell off just 5 percent of another Saudi state-owned behemoth, Saudi Aramco, would have valued that enterprise at more than $2 trillion. Those proceeds could easily finance even Uber's multi-billion-dollar *quarterly* losses for years if needed.

SoftBank's cozy relationship with the Saudis suddenly turned into a liability, after the October 2018 assassination of *Washington Post* columnist Jamal Ahmad Khashoggi was linked to the country's royal family. But while it's unclear whether Son will return to the Gulf for more money—he has raised $14 billion from other sources since the killing—for the foreseeable future the House of Saud will remain the prime beneficiary of this new global traction monopoly's rise.

DURING ITS FIRST DECADE, Uber didn't play nice with cities. Its Greyball program, which ran from 2014 to 2017, scoured user data to identify accounts being used by taxi regulators for sting operations in cities where the ride-hail service was illegal. Once so tagged, Greyball would thwart municipal officials' attempts at enforcement by spoofing their screens with phantom cabs. But in 2018, the company attempted to turn over a new leaf. Under new leadership, Uber extended an olive branch to cities everywhere. "We're ready to do our part to help cities that want to put in place smart policies to tackle congestion," wrote Uber CEO Dara Khosrowshahi in a blog post. "Even if that means paying money out of our own pocket to pass a tax on our core business."

Uber's embrace of congestion pricing was initially well received, coming as it did on the heels of maverick founder Travis Kalanick's ouster the year before. But I didn't buy it. I was already on edge about congestion pricing's prospects in an age of algorithmic financial fudgery, and I didn't trust Uber. I wondered—was congestion pricing really a tax on Uber? Or was the company simply borrowing another trick from its Gilded Age predecessors and gearing up to collude with city governments to clear the streets of competition?

The plot seems too diabolical to be true. Uber's surface motivations for backing congestion pricing make sense. Its cars suffer as much from slow traffic as everyone else—measurably so, even, as lower speeds generate fewer revenue-miles. In the short run, Uber will have to eat the cost of congestion tolls. Here's the kicker, though—so will everyone else. The fees will speed the burn rate for Lyft, too, hastening its inevitable demise. Soon enough, Uber could pull a switcheroo, and start passing on congestion charges directly to you and me.

Meanwhile, Uber would also gain an upper hand on local government. In the downtowns where its fleet has grown the most, the company already has the power to pile on traffic at will. As its market domination became complete, it would achieve near-total control over not only the prevailing level of congestion but how much revenue cities earned as a result. Treat the company well, and mayors might reap the optimized windfall of streets topped off with private cabs around the clock. Treat the company badly, and it might push steep discounts for shared rides into congestion zones to reduce the flow of tolls to a trickle. Sure, your city's streets would be clear for buses to run free—you just wouldn't have the money to keep the system afloat. Something would have to give.

In the streetcar era, nascent traction monopolies cut deals with cities to consolidate their control over urban mobility. In Philadelphia, when the streetcar syndicate's Union Traction Company absorbed the last of its competitors (including the tragically named People's Traction Company), it agreed to repave every street where its tracks ran. Five-cent nickel fares, set under an earlier agreement, were locked in place for years to come. Similarly, cities might simply grant a ride-hail monopoly in return for cash or other guarantees. If Uber and its ilk fall on hard times, they might simply assume municipal control over the services—much as happened when streetcar systems started to decline after World War I.

Despite the stability its city-sanctioned traction monopoly provided, Philadelphia paid dearly. As trailblazing *New York Evening Post* journalist Lincoln Steffens wrote, under the traction monopoly Philadelphia was both "corrupt and contented." (Perhaps the Big Apple rag's editors were just jealous. When Jay Gould's Manhattan Railway Company consolidated the city's elevated railroads in the early 1880s, he'd doubled the price of a ride to 10 cents.) Unlike New York, where—under the boot of an equally powerful traction monopoly—proponents of subways persevered, bringing the city into the twentieth century with speed, the City of Brotherly Love failed to modernize en masse. Today, it still maintains one of North America's largest functioning streetcar networks. Its subway system, however, is woefully undersized. The unwillingness, or inability, to deal with Widener's gang left Philadelphia at a severe economic disadvantage even as other cities zoomed ahead.

Capitulation was not the only option. Elsewhere, the threat of the traction monopolies was met not with new technology but with political force. In 1905, Seattle voters approved the creation of a public power utility to compete head-on with the Puget Sound Traction, Light & Power Company. After a bitter struggle, the public authority eventually took over Puget Sound's streetcar lines, buying the company out in 1918 amid a wave of wartime nationalization. The old streetcar system, converted to "trackless trolley" buses powered by overhead wires—and by City Light, a separate entity—operates to this day under municipal ownership.

Money Talks

Earlier we saw how the term *autonomy* can draw our attention away from the vital role of government in providing basic infrastructure, to focus our fascination on technology and automotive products instead. In the years to come we'll face a similar struggle to clearly see the new flows of money unleashed by AVs. Financialization entails a web of destabilizing shifts. It rewires economies, companies, and even families around the measures, mechanisms, and meanings of banks, markets, and regulators. It does most of its work under the radar, though. And urban mobility is now falling into its grasp.

The huge risks of this marriage between high tech and high finance are clear. Uber's 2019 IPO filings revealed—despite the company's nice talk about congestion pricing a year earlier—its true intentions toward cities all along. Public transit was the competition, and Uber planned to win. What's more, instead of using financial engineering to produce socially beneficial outcomes—like surge pricing's (supposed) supply-inducing effects—the company had shifted instead to a sophisticated new form of microtargeted "dynamic pricing." By mining data on current traffic, your ride history, the weather, and even the remaining charge on your phone's battery, the company showed its eagerness to deploy predatory pricing with little regard for unintended consequences.

We can police the bad behavior of a single company. But Uber's betrayal of cities is only the beginning. Whether it's a traction monopoly, secondary markets built atop MaaS platforms, or bad behavior unleashed by hyperconnected tolling—ever-more-effective methods to monetize our highways, streets, and even sidewalks are on the way. But how far will the financialization of mobility go?

What's clear is that the creeping normalization of market logic won't stop at merely implementing mobility policy, as congestion pricing is seen to be doing today. It will become a tool for actually making policy, with powerful interests, rather than people, pulling the levers. When access to road space is a winner-take-all auction pitting bot against bot, commuter vans will end up in costly competition with computerized chauffeurs. High-value-goods carriers will outbid mass-transit movers of "low-value" people. Those without bank accounts or good credit will find themselves without a ride. How long will it be before any remaining illusion of fairness—about who gets to move where, when, or how—disappears entirely?

We may also set ourselves up for another financial crisis, this time in mobility markets. Much like the toxic housing debt that helped trigger the global financial crisis of 2007, the increasingly predictable revenue streams of automated mobility services will be ideal for conversion to asset-backed securities, which already "have been created based on cash flows from movie revenues, royalty payments, aircraft leases and solar photovoltaics." Unleashed into secondary markets, these algorithmically traded financial instruments could do unspeakable damage. A rogue outfit like Enron, eager to corner the market, could withhold the supply of transportation to goose the value of the securities it holds. Savvy speculators might trade algorithmically with stealthy precision, evading detection, manipulating the supply of mobility in undetectable, deleterious ways.

This is about far more than control of city streets. These scenarios warn of a mind-numbing realignment of money and power. Consider the case of CalPers, the pension fund for California public employees. With some $300 billion in assets, CalPers is one of Wall Street's biggest institutional investors. Future investments in traction monopolies may pay handsomely, and the fund is already deep into ride-hail through its various holdings. But as the group's 1.6 million shareholders age, conflicts of interest are unavoidable. Consider, for instance, the interplay between mobility, housing, and health care. Facing insurmountable care costs for unprecedented numbers of seniors, national governments are eager to encourage people to age in place. Taxibots will provide a crucial connection for homebound seniors to access care and companionship. For pension funds like CalPers, this seems like a win-win. It makes money on its investments in traction monopolies, and fosters an innovation that improves its beneficiaries' lives and eases pressure on public finances. The alternative—allowing the housing market to collapse as retirees move en masse into care centers—is unthinkable. But what happens when such policies turn a traction monopoly into a vital service deemed "too big to fail"? Will health-system savings simply be gobbled by a rent-seeking traction company? Aside from the ethical concerns, the numbers involved in this web of money are so enormous that the future solvency of the welfare state may well hang in the balance.

Even as tomorrow's global traction monopolies take shape, we don't yet know where their ambitions end. Google set out a mere two decades ago "to organize all the world's information." In that regard, it has surpassed everyone's wildest expectations. Now, through ventures such as Waymo, it seems intent on organizing the entire physical world, too. Today, Amazon's body-tracking technology spies on potential shoplifters in its Amazon Go cashierless stores and would-be shirkers in its warehouses, measuring the movement of heads, hands, and feet in stunning detail. In the future, these surveillance systems might be turned on pedestrians throughout the world's cities, mapping their most minuscule movements in order to monetize them all. But if this is a bridge too far . . . where will we draw the line? Will we be both corrupted and content, or will we choose a better path?

8 Urban Machines

If enough people see the machine you won't have to convince them to architect cities around it. It'll just happen.

—Steve Jobs, cofounder of Apple, on the Segway scooter (2001)

I t may sound to you like ancient history, but as recently as 1945 most Americans still lived in cities and rode trains and buses to work. That was the year my parents were born in the West Oak Lane section of Philadelphia. But by the time I came into this world, in 1973, their neighborhood—like so many others around the country—had been bled out by suburban highways. Cars carried millions away from aging downtowns into vast new regions. In our case, the exodus stretched all the way to Maryland, where I was born.

The automobile is a powerful engine of dispersal, a *centrifugal* force that splays the population out across the land. It shaped America in the twentieth century, and reshaped the UK and Europe, too. Today, China is in the thick of its own auto-enabled urban expansion. The Middle Kingdom adds some 20 million new cars every year to its cities, which it encircles with ring roads. Beijing is already on its seventh.

The driverless revolution will undo all that. Specialized AVs will deliver better services, but only to those dwelling within their limited footprints. You'll want to live close in, underneath their umbrella of automated convenience. Similarly cheap, continuous delivery will put a new premium on proximity to distribution hubs—centers of commerce that will draw businesses, jobs, and residents much as the village well once did. And financialization's ruthless imposition of market forces on travel decisions will encourage clustering with a stronger hand than any government policy. Together, these forces will exert a *centripetal* pull, drawing us to the center. Unlike the car culture of today, AV-enabled living won't disperse us. Autonomy—the technology of machine independence—will bring us together instead.

This won't be the first time that automation has strengthened the center. Throughout urban history, we've used self-tending machines to overcome barriers to urban growth. The push-button elevator, introduced by Otis Elevator Company in 1894, did away with the expense of lift operators, accelerating the vertical expansion of cities. The automatic telephone exchange, invented in 1892 by an undertaker in Kansas City, helped municipal governments coordinate police, fire, and sanitation as the population, area, and pace of industrial cities expanded with breathtaking speed. And when electric streetcars and motorized vehicles mixed it up with horse-drawn ones, automatic traffic signals (first patented in 1917) unclogged traffic-snarled streets. All of these inventions had a singular effect. They allowed more people to share the same urban space safely and efficiently. Our cities simply couldn't function without them. We invent technologies to automate cities because *we have to*.

Today's digital technologies are no different. Like their mechanical and analog predecessors, they are helping us manage the problems of unprecedented urban growth—the number of city dwellers worldwide expanded from under one billion in 1950 to more than four billion in 2020. The invention of AVs is simply the next act in this long-running story.

THE REVERSAL FROM car-powered dispersal to a new era of computerized concentration raises many questions. In the pages that follow, we'll explore how AVs will help us overcome today's obstacles to urban growth by expanding in more sustainable ways. We'll see how these technologies will allow us to use urban space far more intensively than ever before, and we'll see the extraordinary value this will create. Along the way, we'll also discover valuable lessons on how we might tame the AV's potential excesses.

Our journey will take us across an imagined city of the future, from its center to the hinterlands, following a path that urban planners call a *transect*. A transect is a cross section of a city and its surroundings, sliced open to reveal its structure, a little like reading the growth rings on a tree. In that sense, this study grounds us in history by revealing how cities developed over time. But the transect also helps us sketch the road ahead, providing a canvas on which to plot our own ideal future. It helps us think about where buildings should be located, the amount and kind of open space we need, and what transportation links are required for people and stuff to move around.

Our transect consists of four zones (Figure 8-1). In the urban *core*, the centripetal pull is strongest—AVs of all shapes and sizes are unleashing a flurry of movement and construction that creates intense concentrations of human activity. Surrounding this hub is the *fulfillment zone*, the twenty-first century's industrial district, where people and intelligent machines work side by side to make, remake, sell, and move a mountain of goods. Farther out, *microsprawl* reorganizes the suburbs, bringing lots of new, affordable, medium-density housing within easy reach of mass transit. Finally, in the metropolitan fringe, a vast place called *desakota*, the countryside is replaced by a chaotic, periurban web of automated agriculture and industry.

This transect is not a blueprint for communities to copy. It is a teaching tool, and a whiteboard for thinking about how the three big stories of

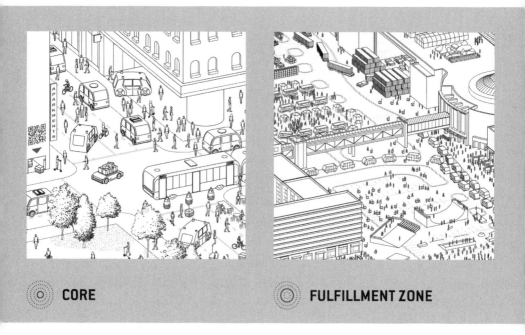

CORE   FULFILLMENT ZONE

the driverless revolution could play out in the coming years. To use it most effectively, first think about what's happening in each zone of the transect. What goes on here? Where are AVs and new mobility services moving in? What desirable and undesirable impacts do they have on how people live, work, and play? Only then consider possible responses. What does government need to do, or incentivize others to do, differently, to adapt?

In the pages ahead you'll find the basic materials you need to start crafting some designs of your own. And that is the first step toward taming the AV.

Rebuilding the Core

I'm often dumbstruck by the lengths to which we go to store cars. Take 200 Eleventh Avenue, a new luxury condo tower that opened in Manhat-

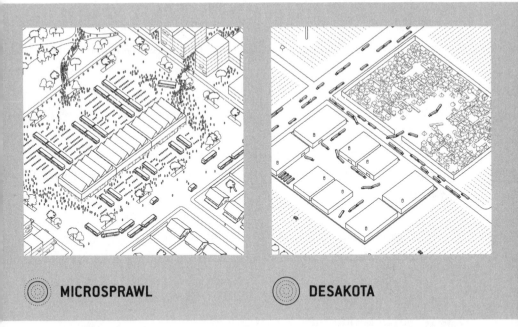

MICROSPRAWL **DESAKOTA**

Figure 8-1. *Transect of the driverless city.* The transect reveals long-term changes in urban regions in the driverless future, from the center (core) to the periphery (desakota).

DASH MARSHALL.

tan a few years back. The stand-out feature? A "sky garage" car elevator opening directly into each $20 million apartment.

But the most impressive parking garage I've ever seen was dug deep into the bedrock of the Bosphorus. In 2013, as workers put the finishing touches on Istanbul's newest civic monument, the Supreme Courthouse, I sneaked a look at a subterranean space that would soon become one of the most secure in the Middle East. Feeling a bit like Dante in the opening stanzas of the *Inferno*, I followed the structure's designers down a gently twisting helix of concrete, into the depths below.

Like the ancient city, Alexis and Murat Şanal are a union of East and West (she's an American, he's a Turk). And as part of an architectural avant-garde in Istanbul, they struggle constantly to reconcile the city's unparalleled history and its simmering tensions. Even here, in a secret space where no critic's camera will ever pass, they've embedded powerful symbols. Every few meters, small chambers branch off the main loop—which I take as a subtle stab at the clannish power structure of Turkish society. More obvious is a continuous strip of bright white LEDs that light the way overhead from top to bottom—unmistakably reminding all who pass of the sacred continuity of the rule of law.

Not all parking decks are so poignant. But this one might just be the pinnacle of the form, because the driverless revolution will make such wasteful excavations and erections obsolete.

WHERE DO CARS GO AT NIGHT? presents itself as a bedtime story for the AV age. "Cars don't need parking garages anymore," the tale begins. Since these insomniac vehicles never rest, we learn, they'll store themselves in perpetual motion instead.

This story isn't quite *Alice in Wonderland*, but it is full of supernatural imagery. Drawings conjure up images of sprightly, civic-minded autos that "clean the streets . . . take care of plants in the park . . . and . . . repair broken things in the city." Produced by Daimler's Moovel Lab in Stuttgart, Germany, it does bring some creativity to what would otherwise be a dull tale of municipal housekeeping. And it's a pretty soft sell of autonomy, sending kids off to dreamland with visions of a driverless utopia.

The tales we tell adults about urban change in the age of automation, however, are less subtle. "As fracking upended the oil industry by giving new life to old fields," reported Bloomberg News breathlessly, "so the driverless future offers to free up whole new neighborhoods." The analogy is intriguing. It makes me imagine some of my urban-planner friends mixing it up over cocktails with the wildcats of the real fracking business. Maybe they'd tie one on together, head downtown, and drop a wellhead square in the center of a daily lot.

It's hard to overstate how much urban planners despise parking.

Parking lays fallow astronomical amounts of cities' most valuable land, and sucks the life out of neighborhoods for blocks in every direction, too. But what if this toxic waste could be turned into an asset? Gensler, a global design firm, estimates there are nearly six billion parking stalls worldwide—500 million in America alone. (More ambitious estimates put the US number at nearly two billion.) Most US downtowns have so much land devoted to parking, you could fit a whole new set of buildings in between the existing ones. Even Manhattan, the least car-friendly place in North America, has more than 100,000 public off-street parking spaces. The situation in the UK isn't nearly as bad. Yet one study identified more than 10,000 reclaimable parking lots where homes for as many as one million people could be built.

Real estate developers hate parking, too, and it is their distaste for it that will ultimately reshape cities, regardless of what urban planners think or do. In fact, developers have long locked horns with local governments over minimum parking requirements—a relic of a fading era when a planner's job was to make sure developers provided *enough* off-street parking for new homes and businesses. Laws mandating the construction of parking facilities increase costs while also taking away space that could be used to build more profit-producing units. Today, many communities are phasing the obsolete rules out and putting parking *maximums* in their place to encourage walking, biking, and transit use instead. Builders support such reforms because doing so boosts their bottom line. Hence the excitement about AVs and the potential parking diet they'll bring on.

The trouble is the transition. Tomorrow's buildings are already on the drawing board. Yet many developers are looking 30 years ahead and realizing that parking garages will be obsolete before their construction loans are fully paid off. So while, in the future, we may require a tiny fraction of the parking we do today, how do you provide the space needed for cars now, while reserving the option to take it back for housing or open space at a later date?

As it turns out, it's harder than it looks to build a parking garage that's suitable for conversion to housing, office space, or a logistics hub down the road. A structure's layout must be flexible enough to accommodate unpredictable swings in the market and technology. But it must also still be economical enough to build today and retrofit a decade or two in the future. The task is made all the more difficult because the design tricks used to cut corners today—sloped floors, low ceilings, elimination of risers for plumbing and cables, and elevators positioned at the corners rather than in the center—make such structures unsuitable for human occupancy. Merely raising the ceilings back to a height that people find acceptable increases a parking structure's construction costs by 15 to 20 percent. More substantial design changes—like moving ramps to the exterior of the structure, or relocating support columns so they don't block your view when condos replace cars in 2050—cost much more.

None of this has stopped developers from trying, and high housing demand has spurred some successful parking garage conversions. Take Knightley's, a futuristic parking garage built in 1949 on the edge of Wichita, Kansas. When it opened for the first time, downtown shoppers could drop their car with a valet before heading off to nearby stores. Purchases would be delivered by courier, hoisted by dumbwaiter, and deposited in an air-conditioned lounge—for later pickup along with the family car. There was even a safe for valuables. But when Knightley's reopened in 2018 after a two-year renovation, 40 luxury apartments welcomed a new generation of shoppers more likely to be serviced by Amazon Prime robovans and Starship conveyors instead.

While the fracking of parking was little more than a clickbait lede, the more I consider the metaphor, the clearer the parallels come into focus. First, the technologies that made fracking possible—better maps (of underground geology) and better technology for manipulating the physical world (horizontal drilling)—are in the same class of innovations that are bringing AVs into being. Second, while building garages for a postparking future will make sense in cities where land values are high and ride-hailing is widespread, just as with fracking, the big money lies in virgin territory—surface parking. Los Angeles County alone has nearly 19 million parking spots consuming 200 square miles of land, with an estimated market value of $350 *billion*. Finally, like fracking, urban real estate is a highly speculative, volatile, debt-fueled business. The boom could come on at any moment, with startling speed.

⸻

THE ARRIVAL OF AVs will hit big-city downtowns first and hardest (Figure 8-2). And this looming threat has already prompted dozens of cities, from Toronto to Taipei, to develop their own visions for how the driverless revolution should play out. These mayors and their teams, often in consultation with residents and businesses, are trying to understand how they can spur appropriate uses of AVs while also preempting destructive ones. They're dealing with issues as diverse as aging, employment, housing, and watershed management—all are subject to the destabilizing effects of automated mobility. Given this wide purview, it bodes ominously that the clearest sketch of city governments' collective thinking—the *Blueprint for Autonomous Urbanism* published in 2017 by the National Association of City Transportation Officials, a club of city-street czars—focuses solely on the fundamentals of future streets.

The *Blueprint* is at once remarkable and curious, an overly ambitious attempt to draft precise patterns for a technology that's still fresh off the drawing board. But it fills an important gap in foresight about AVs and cities. Building transit systems and changing the setup of streets can take many years. These choices, once made, shape the city for decades afterward. Municipal leaders must think far ahead, however foggy their understanding of how the technology, economy, and society of the future may turn out. The *Blueprint* gives cities detailed guidelines for taming any AV that might roll up, using their existing knowledge, legal powers, and purse strings.

But the *Blueprint*'s most intriguing ideas aren't about street design at all. Instead they deal with software design. The majority of its recommendations deal directly with the computer code that will shape the "dynamics of future streets" as much as any physical changes do. For instance, one-way streets—which are considered outdated because they limit street-network flexibility and encourage speeding—are abandoned in favor of software-synchronized two-way traffic. Another proposal calls for exploiting computer-controlled steering to paint narrower travel lanes, and eliminating curbs to make walking easier. New layers of data and code orchestrated by city authorities would manage all kinds of movements—such as directing vehicles to curbside spots for passenger and goods pickups, closing streets to traffic, and even setting the spacing behind moving vehicles to ensure pedestrians have enough time to cross between.

The *Blueprint*'s most important contribution, however, is the pecking order it establishes between people and machines, with humankind placed on top. Ghost roads are banned, unequivocally. "AV-only lanes should be discouraged," it advises. And "streets of all sizes are designed for people, not vehicles." To underscore this point, the guidelines insist that pedestrians should never be required to carry or wear any type of beacon to be detected by AVs—simply being human ought to guarantee one's right to exist in safety.

The *Blueprint*'s goal is to show cities how to draw a defensive perimeter against the potential excesses of autonomy in city centers. Many city transportation officials see AVs as a threat to the hard-fought gains in human-centered mobility of the last 20 years—huge increases in cycling, pedestrianized plazas, and reduced speed limits. The *Blueprint* urges mayors to consolidate those gains while also chalking out a sensible place for automation. It shows them that it is possible to exploit AVs to boost transit capacity and the convenience of taxis, while also expanding space

Figure 8-2. *Urban core.* In the urban core, automation moves more people and goods than ever, creating more productive and lively spaces with little need or room for cars.

DASH MARSHALL.

APARKMENTS

for walking and biking. And it explains how software will let us govern the intermingling of people and self-driving machines on streets and sidewalks with remarkable precision. While the blueprint's name tantalizes us with false visions of a finished design, as a template for thinking about how AVs can serve cities—rather than the other way around—it will be invaluable in the strange days ahead.

Migrating to the Fulfillment Zone

Traveling along the transect out from the downtown core, we enter a fast-changing ring of the city—the *fulfillment zone* (Figure 8-3). While it is every bit as titillating, don't confuse it with a red-light district. The fulfillment zone is the twenty-first century's loft neighborhood, those indispensable urban districts that thrive on lower-than-average rent and better-than-elsewhere transportation.

The fulfillment zone's contours are shaped by continuous delivery—buildings are of secondary concern to the robot-powered work at hand. Any kind of structure will do, and every city will find a different type—vacant shopping malls, obsolete hospitals, or a pile of disused shipping containers.* In between them, the streets are abuzz with conveyors and mules, endlessly shuttling goods to and fro.

While automated commerce is king here, the dynamism it produces also attracts an extreme mix of uses. Everywhere, buildings sprout a rich assortment of enterprises and livelihoods supported by self-driving machines. But this isn't just conspicuous consumption. This is a hyper-circular economic cluster, where value chains stack up in tightly choreographed loops of recovery and upcycling. The fulfillment zone is also a place of community. People come and go on one-wheelers and 300-

---

* For instance, in Pontiac, Michigan, the site of the Silverdome, demolished in 2017, will become the home of a $250 million Amazon distribution hub. Omar Abdel-Baqui, "Pontiac Mayor: Amazon May Bring 1,500 Jobs to Silverdome Site," *Detroit Free Press*, September 19, 2019.

passenger software trains alike. Learning and leisure are everywhere, and are often indistinguishable to the untrained eye.

The fulfillment zone's magic arises thanks to three converging migrations—two involving goods, one involving people, all of them mobilized by AVs.

First, this is the beachhead for online retailers staging same-day delivery incursions into dense, affluent urban markets. The fulfillment zone is closer to the core than the big regional distribution centers built just a few years ago. But it is not so close as to compete with office blocks and luxury high-rises for land. This is where the complex equation of rent, transport costs, and travel time reaches an efficient equilibrium. As one official at Prologis, a real estate trust that runs distribution centers, explains, "rental costs account for just 5 percent of our customers' overall supply chain costs, while transportation accounts for over 50 percent. . . . Customers are willing to pay higher rents if it means they can reduce transportation costs and be closer to major consumer markets."

A second migration consists of small retailers and service providers fleeing the high costs of traditional downtown storefronts. Already today, ghost restaurants have relocated to delivery-driven kitchen clusters in the fulfillment zone. Many more will follow, hollowing out the core of even the most hallowed culinary centers. After all, even though Manhattanites and Parisians bear some of the highest housing costs in the world, presumably to be close to fancy shops and fine dining, according to a 2018 survey they're eating out less than ever, thanks to delivery apps.

The ghost-business boom will reach far beyond food, however, as cheap AVs widen the range of retail disruption. Services will set up shop in the ample quarters of the fulfillment zones, and simply fetch their clients by taxibot or driverless shuttle. Conversely, online-to-offline services— already one of China's fastest-growing sectors today—may exploit AVs to deliver mobile tutors, hair salons, and dog groomers on demand from their base of operations in the fulfillment zone. Finally, the public realm will be a place for experimenting with new forms of AV-enabled retail—

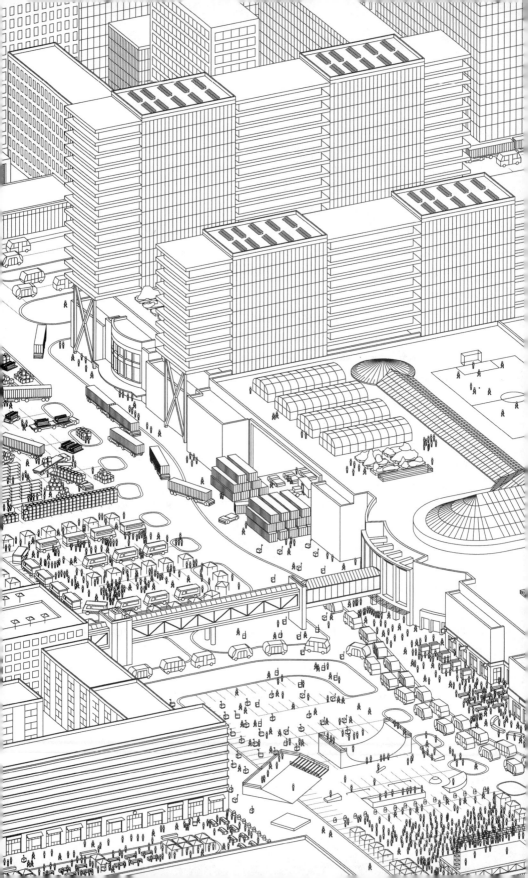

mules and caravans will power extreme pop-up shops, delivering just-in-time goods to self-serve street corners.

The third group decamping to the fulfillment zone is looking for homes. But this kind of city living will force developers to get creative with mixed-use designs. In London, for instance, in recent years a rapid expansion in the number of same-day-delivery terminals has bumped up against intense demand for housing. With a shrinking number of suitable parcels left to build on, the two uses are now in direct competition for the same land. But recent projects in the British capital have combined more than 1,000 flats and a half-million square feet of last-mile logistics facilities. The clever concept's name—"sheds with beds."

Today, it's difficult to imagine how we'll reconcile the conflicts created by automated commerce and urban living. But *sheds with beds* highlights the design possibilities to come as the fulfillment zone creates jarring new juxtapositions of use. Imagine how it might play out. In the predawn hours a sortie of mules silently sallies forth, hauling parcels a short distance into the urban core. Not far behind, commuters trickle down from the towers above, picking from a swarm of idling rovers to scoot off to work. During the day, conveyors come and go on their on-demand runs. Vertical warehouses were common in the nineteenth century—reimagined with residences stacked up into the sky overhead, they could quickly rise again across the world's great cities.

Throughout history, the nexus of transportation and retail has been a rich generator of experimental designs that go on to capture new commercial niches—station hotels, shopping malls, and modern airport terminals come to mind. And so, as automation upends the movement of goods in cities, the fulfillment zone is where tomorrow's most interesting architectural experiments may take shape.

Figure 8-3. *Fulfillment zone.* A mash-up of distribution hubs, ghost restaurants, pop-up retail, and residential spaces, all connected to the surrounding region by AV shuttles and continuous delivery. DASH MARSHALL.

Microsprawl

As we continue our journey, we enter the vast residential expanses where most of us live. Here, a new combination of automated mass transit and single-passenger self-driving vehicles will give rise to a new neighborhood pattern I call *microsprawl*. Unlike old-fashioned suburban sprawl, microsprawl has the potential to both increase the supply of affordable housing and reduce carbon emissions. But realizing these benefits will require us to rethink current assumptions about transit, walkability, and good urban design.

The first concept that will need a major overhaul is what urban planners call *transit-oriented development*, or TOD for short. TOD revives the way things were done before the car arrived. And it's a straightforward idea—build a train, and then put up buildings nearby, ideally within walking distance. This form never went out of style in Europe, and is widely deployed throughout Asia. But the practice took some time to come back into fashion in North America, beginning in the 1990s. Today, most TOD schemes use zoning laws to incentivize developers to build taller and closer together than in the past. The hoped-for outcome is that people living, working, and visiting the new district can meet most of their daily needs on foot and hop on a train or bus for anything else, without needing to own a car. TOD creates a positive set of feedback loops, meant to unwind the damage of car-style sprawl. It creates access and mobility with less energy, and expands the supply of housing, theoretically lowering prices.[*]

---

[*] While there has long been a consensus among urban-housing experts that increasing the supply of housing lowers prices, recent research has challenged that view. See, for example, Andrés Rodríguez-Pose and Michael Storper, "Housing, Urban Growth and Inequalities: The Limits to Deregulation and Upzoning in Reducing Economic and Spatial Inequality," *Urban Studies* (September 2019), https://doi.org/10.1177/0042098019859458.

Figure 8-4. *Microsprawl.* An improvement over its low-density predecessor (bottom half), microsprawl (upper half) is defined by large tracts of higher-density housing accessed by personal rovers and software-train mass transit. DASH MARSHALL

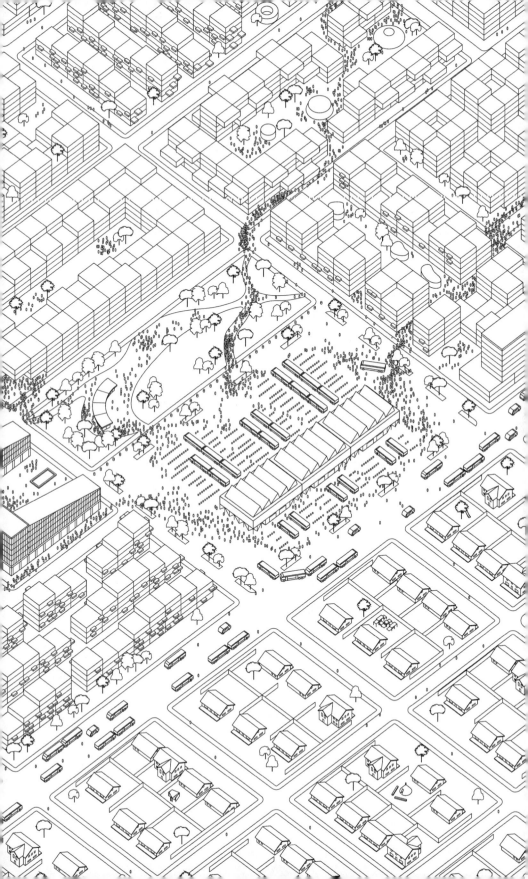

At first glance, automation looks like a force multiplier for transit-oriented development. Prominent urban planners like Peter Calthorpe, a founder of the Congress for the New Urbanism, have embraced the synergy of software trains and compact neighborhoods. It's easy to imagine driverless shuttles playing a role in linking satellite neighborhoods designed without cars in mind—boasting more footpaths and less parking, for instance. That's the thinking behind Switzerland's Les Vergers Ecoquartier, a 1,200-unit sustainable housing development at the edge of Geneva that's connected to downtown transit by robovans.

But the smaller AVs get, the more complicated and counterintuitive their impacts on neighborhoods will become. Scooter- and bike-share services, for instance, have spread quickly in the last few years. *Micromobility*, as this emerging sector is called, aims to serve the under-five-mile journeys that make up 60 percent of all trips in the US. Urban planners have mostly welcomed the arrival of these new services, expecting that they'll provide an alternative to the congestion and carbon emissions of private cars. Indeed, the biggest immediate threat they pose is to the ride-hail industry itself. In Sacramento, California, for instance, Uber now provides more rides via its Jump bike-share platform than via ride-hail. The only problem—and it's a big one—is that micromobility appears to be "disrupting" walking, too, to borrow a Silicon Valley euphemism. Early surveys indicate that up to 37 percent of scooter trips replace walking journeys rather than ones taken by private car. Historically, walking is an inextricable part of what makes TOD work. It's the key to everything. Do cities need to start worrying about scooter-powered sprawl?

Perhaps. But is there a way to turn this threat into an opportunity? Can we couple two seemingly contradictory impacts of automation—the improvement of transit and the subversion of walking—to mobilize a new kind of neighborhood? Could rovers actually help us improve TOD instead of abandoning it?

DURING THE HOUSING BUBBLE of the early 2000s, "Drive until you qualify" for a mortgage became a do-or-die mantra for first-time homebuyers. In a city of rovers, the same might be true, but you'd hop onto a self-driving scooter instead of sliding into an SUV.

To understand why, we need to think geometrically. A widely used rule of thumb in urban planning assumes that people are willing to walk only about 20 minutes to a transit stop—in practice that works out to about a one-mile walkshed for TOD projects. Cruising along on a rover, however, the same person can cover more ground in the same period of time—as much as *five* miles, circumscribing a zone some 25 times larger in area (remember, $A = \pi r$). This new, vastly larger transit-oriented territory can, in theory, be developed at the same density as the older, smaller one. And in so doing, we've radically expanded the supply of land suitable for housing that's well served by transit, allowing people to live with fewer cars.

Alone, this strategy should drive the cost of housing down and reduce carbon emissions. But we don't have to stop here. We can also use automation to boost capacity on the main transit line serving the center of our new mobility district. This makes it possible to also up-zone the whole neighborhood for higher-density development than traditional TOD. We can build townhomes and apartment buildings where single-family homes might dominate today. By expanding horizontally with rover-based mobility, and vertically through denser, taller residential development, we might be able to bring 25, 50, or even 100 times more dwelling units online within reach of that one stop. And we could do it with highly automated buses rather than trains, reducing construction costs. That would make it easier for cities with limited finances to build mass transit and support lower fares that attract more riders.

Microsprawl's potential to expand TOD's ambitions could help many communities break through a current deadlock. Demand for housing has outstripped the supply of available sites to build on, transit systems can't bear the burden of more TOD along existing lines, and opposition to increased density in existing communities is widespread. Microsprawl would offer a new tool, allowing planners to spread high-density infill housing over a larger area, giving them more flexibility to avoid conflicts. Automating core transit lines with software trains as well as more conventional communications-based train control (CBTC) technology would allow them to absorb the additional strain. Meanwhile, moving more local traffic out of cars and onto rovers would reduce emissions and take large vehicles off the road. All of this can be achieved by simply increasing the range within which people are comfortable traveling to and from a transit station.

Microsprawl will also create enormous challenges. The collateral damage to walking and public spaces may be severe. Tomorrow's microsprawling neighborhoods will feel very weird—but it's hard to know which of the good and bad characteristics of today's walkable hubs and yesterday's car-centric communities they'll inherit. Smaller downtowns and village centers that already struggle to attract crowds may simply wither away, as rovers replace walking. Meanwhile, the same paths built for rovers will provide a distribution network for conveyors, accelerating the creative destruction wrought by continuous delivery on local shops and restaurants. Here's the thing, though—achieving a massive increase in affordable housing and much-needed reductions in transportation-related carbon emissions may be worth sacrificing a few downtowns. The urgency of climate-change adaptation may present such gut-wrenching choices much sooner than we think.

Still, I suspect we'll quickly adapt and find ways to exploit the new capabilities of rovers. You'll summon them from phones and wearables, or even by saying their name out loud to a smart streetlight. They'll learn our habits and pre-position themselves at popular pick-up spots like transit stops, restaurants, and schools. When not in use they'll run off and hide away down back alleys, on side streets, and in old garages. And once we come to understand rovers' novel capabilities for human-centered mobility, we'll find endless opportunities to regroup around a new set of gathering spots within the microsprawl. These vast new neighborhoods crisscrossed by herds of digital steeds may feel weird, but so did riding in trains underground at first. When it comes time to pay the mortgage or the rent, we are likely to agree it was worth what we left behind.

Deliver Us from Desakota

The last leg of our trip traverses the blurring boundary between the city and the countryside. At the outer orbits of the metropolitan constellation, a few dozen miles from the urban core, the driverless revolution's gravity is much less intense. Here, the limitless attention span of AVs opens up vast territory. Ghost roads trace a vast web that stretches from horizon to horizon.

This is the most confusing zone of our transect, and it demands a special name. What we find here defies our derogatory labels for the periphery, like *exurb*. Those words describe what this place is not, not what it is. I call it *desakota* instead, an expression borrowed from the other side of the world, a mash-up of the Indonesian words for village (*desa*) and city (*kota*). I first heard it more than 20 years ago in Singapore, from the Kiwi geographer Terry McGee, the term's originator.

At the turn of the millennium, the vast periurban regions surrounding cities like Jakarta, Manila, and Bangkok defied most scholarly efforts at understanding. These were landscapes shaped by global webs of material, migrants, and money that reached in with ease. Modern infrastructure punched through farmlands still tilled with hand tools. An ancient fishing village stood next to a brand-new factory. Rural refugees crammed into mass-produced housing, where they stayed up late toiling away at piecework for overseas retail chains—sewing garments and assembling electronics. A shopping mall would rise above a rice paddy. These places were, it seemed, *all* juxtapositions. But with this single word—*desakota*—McGee captured this colossal alienation and the chaotic coherence that bound it together.

The term appeals to me, and as I contemplated what might come among the hinterlands at the end of the ghost road, it seemed to fit. All over the world, despite its exotic origins, desakota is no longer a far-off place. It is now the way we build along the metropolitan fringe everywhere. And as cheap automated mobility floods entire regions, a fractal web of farms, factories, and towns will spread without limit across the land (Figure 8-5).

Distribution will reorganize first. Ports of the future, plied by self-driving trucks, are as likely to be located in Kansas or Kazakhstan as they are in Shenzhen or San Francisco. "Self-driving vehicles have potential to significantly expand the daily coverage that trucks can provide," argue analysts at commercial real estate brokerage CBRE, "bringing remote storage locations into play." Later, as software trains fan out across the landscape, they'll unleash a new wave of industrial leapfrogging into continental interiors. As Glaeser and Kohlhase dryly concluded in their study of freight costs and cities, "service firms should locate near their suppliers and customers; manufacturing firms should not."

Still, the farms of desakota may be even more thoroughly automated than its depots. Already today, satellites scan every square inch of cultivated terrain, guiding "an armada of sophisticated harvesters so big and expensive that they need to be shared and work 24 hours a day," notes Dutch architect Rem Koolhaas. "In order to feed, maintain, and entertain ever-growing cities, the countryside is becoming a colossal back-of-house, organized with relentless Cartesian rigor."

Figure 8-5. *Desakota.* Far from the region's center, the forces of dispersal start to win out, and cheap automated mobility allows industry to spread deep into automated farmlands.

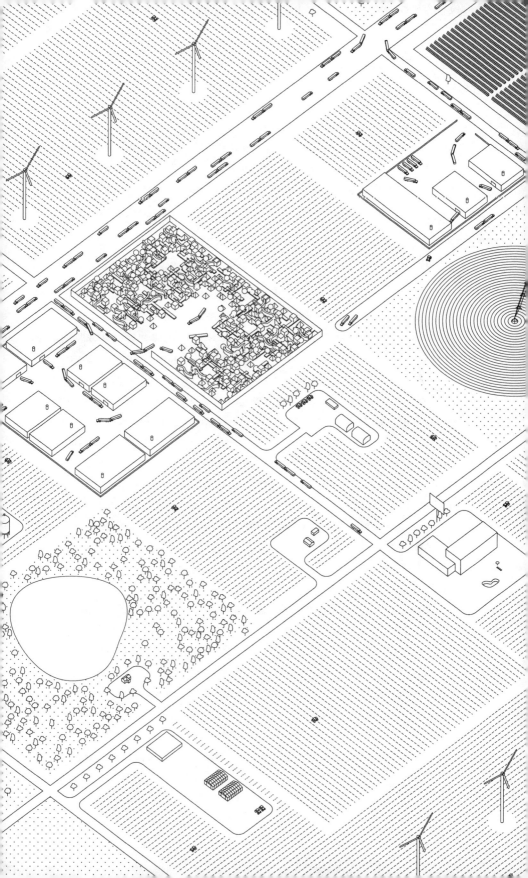

No doubt, people will follow the ghost roads into desakota in search of 21st-century homesteads—both town-tethered supercommuters and sects fleeing the inundated coasts. But once the pheromone trails of the big AVs crisscross the terrain, the countryside's future will be decided. Out here in the extra-urban periphery, a new industrial-agricultural matrix will take shape.

Ghost Road Blues

The transect helps us break up the vast complexity of the metropolis into bite-sized chunks we can get our mind around. But one mistaken message it sends, with all its arbitrary and imagined boundaries, is that separation is a good thing. Nothing could be further from the truth. And there is perhaps no greater risk for how we plan communities in the driverless age.

It doesn't matter what scale you start at. Take streets, for instance. Separating people and vehicles for the last century has forced pedestrians to compete with motorists for scarce public space. In the last 20 years, communities all over the world have started to question that assumption. Many have been steadily reclaiming portions of streets for people rather than machines. As they've done this, walking, biking, and transit have surged. Drivers have been forced to slow down, and road deaths have declined. *Complete streets*, as this approach is known, has worked because it breaks down boundaries. Anyone can use the street, so long as they share it.

But are AVs incompatible with shared streets? Today's AVs devote their full resources to simply spotting other vehicles—and even struggle to meet that challenge. Shared streets, however, create any number of unforeseeable events for AVs to cope with—and engineers are starting to push back. "Rather than building AI to solve the challenge of avoiding a person hopping across the street on a pogo stick," says Drive.AI's Andrew Ng, posing a hypothetical, "we should partner with the government to ask people to be lawful and considerate." In the mid-twentieth century, just this kind of thinking was used to criminalize pedestrian behavior that had, until then, been considered ordinary and lawful. The term *jaywalking*, originally used to describe poor sidewalk etiquette, was weaponized by automobile interests in the mid-twentieth century to criminalize mid-block crossings and expand motorists' claim on streets. Will autonomists like Ng launch a new campaign to stigmatize those who stand in the way of self-driving machines?

Many architects and urban designers seem willing to pave the way for AVs at our expense. Separation is everywhere in today's visions of the self-driving cities of the future. The original master plan for Nanshan Center, a new urban district in the Chinese boomtown of Shenzhen, is a good example. As first drawn by the architects at design powerhouse Kohn Pedersen Fox, it puts AVs on a dedicated elevated platform. (Pedestrians would themselves be channeled to yet another separate guideway on the elevated structure.) While it claims inspiration from New York City's High Line, this seemingly futuristic scheme is as old as the motor car itself—the Regional Plan Association's 1924 vision for New York "argued for dense multilevel traffic and transit solutions that separated levels of rail, wheeled vehicles, and pedestrians with upper-level domains of gracious sun-lit plazas and sheltering loggias" (Figure 8-6).

Zoom out to the neighborhood scale, and separation still haunts the driverless future. Another discarded idea of the modernist past that's being dusted off for the driverless age is the *megablock*—a self-contained neighborhood, tens of thousands of feet in perimeter, free of bisecting roadways and, consequently, cars. Long blamed for increasing segregation and reducing mobility, megablocks feature prominently in AV-ready real estate developments like Singapore's Jurong Lake District and the Milan Science & Innovation Park. Ostensibly, these designs are put forward under the guise of creating vast pedestrian-friendly zones free of cars. But developers like Google's city-building sister company Sidewalk Labs also see them as a tool to keep human drivers away from computerized ones, further simplifying the challenge of rolling out safe AVs. The

Figure 8-6. *A century of separation.* The Regional Plan Association's 1924 traffic-separation scheme for Manhattan. REGIONAL PLAN ASSOCIATION.

company aims to cordon off its proposed Toronto campus—making it "the first place in the world where conventional vehicles will be a thing of the past."

The interface between the old city and these new postcar complexes poses its own set of design challenges. In the seventeenth century, when Amsterdam doubled in size, horse-drawn carriages clogged the streets. "Carriage squares" such as Leidseplein were designated along the outskirts of downtown, providing a place where people could disembark and continue their journey on foot. Similarly, Sidewalk Toronto's designs envision "a small zone that will serve as a transition" with limited parking for conventional vehicles.

And what happens beyond these new city gates? Unlike Enlightenment-era Amsterdam, where pedestrians dominated downtown streets, Sidewalk Toronto's future residents will find themselves face-to-face with AVs. Who (or what) will really rule the roads? The first generation of driverless shuttles, which were designed from the start to operate around people on foot, are programmed to yield—often excessively so. During a recent driverless-shuttle pilot at Hong Kong's waterfront West Kowloon Cultural District, testing was stopped after only a few months. The overly cautious AV was unable to navigate the Chinese city's dense crowds without repeatedly halting and sounding its collision alarm. Local urban-design advocates have called for the "beeping monster" to be moved to a dedicated guideway. But this solution means hard-won space for walking is already being clawed back to make way for robots. The obvious alternative, to dial up AVs' aggression and put pedestrians back on the defensive, could be just as bad.

Other misguided modernist schemes, such as freight tunnels, are getting a fresh look in efforts to design driverless cities. Sidewalk's plan sidelines conveyors into an underground warren, which would do double duty for rubbish removal. Singapore's futurists envision a network of such tubes crisscrossing the entire island nation by 2040. "Purchases are deposited at freight delivery centres located inside the shopping malls, and tagged with the recipient's address. By the time a person reaches home, his goods are waiting at a freight receiving centre near his residential block." And after eventually dropping the idea of elevated AV-only roads for Nanshan, the architects at KPF jumped on the robotunnel bandwagon too.

But underground tunnels for automated freight have been proposed and discarded many times in the postwar era—most notably in 1973, as part of the unbuilt Minnesota Experimental City. Not only do they cost a small fortune to build, freight tunnels lock us into today's traffic-flow forecasts, even as our future shopping habits become less predictable— the tunnels limit the size and shape of vehicles that can be used to move goods and trash, and will create immense security risks. Sometimes it seems that AV freight tunnels would combine all the drawbacks of trains *and* trucks with none of the advantages of either.

Separation is a tempting shortcut, and seems these days to be our brightest designers' and most ambitious developers' first move. Roping off safe spaces for humans avoids the dilemmas created by AVs and people living in close proximity. It also considerably simplifies the engineering of self-driving software. This is a short-sighted tactic, a turn away from the very best ideas of the recent past, like complete streets and transit-oriented development. Instead, it paves ghost roads—spaces dominated by machines where humans fear to pass—right through the heart of tomorrow's new communities.

9 Wrestling with Regulation

> The right to have access to every building in the city by private motor-car in an age when everyone possesses such a vehicle is actually the right to destroy the city.
>
> —Lewis Mumford, 1964

Mayors may have the hardest job in the self-driving future. After the designers, developers, and engineers are long gone, they'll have to pick up the pieces. Yet today, we focus mainly on the national reforms needed to kick-start AV innovation. The real challenge of regulating the impacts of AVs will fall on local governments—as it did in the motor age.

The sweeping scope of opportunities and threats posed by self-driving technology is catching most cities by surprise. But many are now assessing what's in store and their options near and far. One of the first to try to map out its priorities for the driverless revolution was Toronto, Canada's largest city and, through much of the 2010s, North America's fastest-growing tech hub. In 2015, the city commissioned a sweeping study from the University of Toronto, which imagined three "end game" scenarios—one future city where most AVs were privately owned, one where most were taxis, and one where ownership was split. This forecast also provided cities around the world with their first rough guess at the potential safety, economic, and land-use benefits of automation. In June 2019, Toronto followed up this horizon scan with a detailed tactical plan, laying out dozens of specific actions to align AV policy with existing city goals for mobility, equity, sustainability, and privacy.

Since then, other cities have followed Toronto's lead. Working with Bloomberg Philanthropies, I've tracked how more than 135 city governments around the world are gearing up for the driverless revolution. There are many concerns, but three rise to the top of the list of long-term challenges—transit, freight, and public finance. They'll often be mistaken as conflicts over technology, but they'll really be about power—whether the fight is over land, money, or new sources of value such as data. The decisions cities and their mayors make will be shaped by concern for the greater good. But often, they'll be steered off course by narrower political interests too.

Defending Transit

The degree of contempt autonomists hold for public transit is shocking.

Google cofounder Larry Page dreamed of replacing campus buses with driverless cars as far back as his college years at the University of Michigan. In the 2018 book *Autonomy*, a sweeping history of the development of AVs in America, GM's former self-driving-car chief Larry Burns portrays the young computer science student shivering in the frigid midwestern winter, cursing the infrequent shuttle and dreaming of self-driving salvation. "Those who don't have automobiles are forced to ride the bus, which often drives irregularly, and sometimes not at all," Burns writes.

Tesla CEO and cofounder Elon Musk wears his bias on his sleeve. In 2017, speaking to an audience in Long Beach, California, he tore into the idea of collective mobility. "I think public transport is painful. It sucks," Musk ranted. "Why do you want to get on something with a lot of other people, that doesn't leave when you want it to leave, doesn't start where you want it to start, doesn't end where you want it to end?"

It's easy to dismiss these antitransit sentiments as the self-serving ideas of a techno-elite. But as the convenience of ride-hail has spread, the opinion of the traveling public seems to be swinging to their side. Complicating matters is the complex relationship between ride-hail and transit. Many Uber and Lyft trips begin or end at transit stations, a fact the companies highlight often.

But after a decade, a bigger shift is emerging—ride-hail competes with transit everywhere it takes hold. The most comprehensive study available at the time of writing, which analyzed transit and ride-hail ridership in 22 American cities from 2012 to 2018, found that ride-hail's growth accompanied noticeable, and otherwise unexplained, declines in bus (1.7 percent per year) and heavy rail (1.3 percent) ridership. In Los Angeles alone, bus ridership declined 20 percent between 2013 and 2017. What's worse, ride-hail's sapping effect is growing with each passing year. Every new ride-hail vehicle creates more traffic, which slows buses down, too, further discouraging potential transit riders. The death of bus service from competition with ride-hail is becoming a self-fulfilling prophecy.

How can transit systems respond? Should they cut back on service, expand, or change what they do as AV-powered mobility services roll out? We're likely to see all three approaches put to work as transit systems struggle to adapt.

Most transit systems will try to keep doing what they do best—running frequent, scheduled train and bus service along the busiest travel corridors. Rather than try to compete with swarms of taxibots, rovers, and driverless shuttles in the first and last mile, the thinking goes, public resources should be focused on maintaining an efficient trunk system of high-speed mass transit for covering larger distances. A lot of bus systems are already modernizing by straightening routes, increasing the distance between stops, and beefing up off-peak service in areas where there's sufficient demand—such as near universities, hospitals, factories, and nightlife districts. This helps speed up travel and serve more people, making buses more appealing. But automation will also give a boost to beleaguered bus systems. Trials of full-size, driverless city buses underway in Edinburgh and Stockholm hope to both improve safety and radically cut costs, while also offering riders a high-tech experience. In Tsukuba, Japan, transit officials plan to launch their own AV-bus pilot program, which will also feature single-passenger self-driving pods that ferry riders from express AV stops to their front door.

More daring transit agencies may try to transform themselves into *mobility integrators*—expanding into the business of MaaS while continuing to operate trains and buses. This would gradually expand the role of government from simply building and running trains and buses to orchestrating the flow of mobility data and mobility-service transactions. This is the approach of Berlin's BVG, which partnered in 2019 with software firm Trafi to launch the Jelbi MaaS marketplace and app. As the largest deployment of a truly open, publicly managed MaaS marketplace, Jelbi is an important test of whether seamless, side-by-side comparison shopping helps stabilize transit ridership. But there may be a larger end game here. Combine Jelbi with a road-pricing platform like ClearRoad and it is possible to imagine a world where transportation planners could tweak prices for transit and roads simultaneously, not only to reduce congestion but also to encourage low-carbon modes of travel. The establishment of such a regime, a true *transportation utility*, would be a historic shift in how we govern mobility.

Finally, the driverless revolution presents an opportunity for transit systems to radically reimagine the service they provide. Doing so will depend not on introducing new technology, ironically, but on reinventing the transit workforce. While autonomists (mostly men like Page and Musk) boast about the big savings from driverless buses, ask any woman or senior citizen if she relishes the prospect of riding a bus with no uniformed crew, and you're bound to hear a rather different perspective. As the AFL-CIO, a coalition of some 32 labor unions, points out, "the presence of an operator ensures that someone is there to respond to emergencies and summon first responders, prevent unattended buses from becoming magnets for crime, and to provide a backup in the case of technological failure." What's more, the support of organized labor will be essential to the safe, effective rollout of automation itself. If transit automation is seen exclusively as a tool for slashing jobs, it is sure to be blocked by unions— and rightly so.

Instead, future automation efforts in public transit must find ways to deliver exceptional and unique value. That will mean personalized and responsive, customer-centered service. What if the money that goes to pay drivers and dispatchers were spent training a new kind of transit worker to tend to riders' other needs? The bus crew could assist elderly and disabled passengers, sell food and beverages, clean passenger areas, and provide additional security.

Beyond these basic functions, though, transit vehicles could be transformed into pervasive public-service delivery points, extensions of social services, health care, education, and child and senior services. Buses have always served as a kind of "third space," where social connections are formed on the fly. But could they be civic spaces in the self-driving future, too?

Finding Room for Freight

How will communities manage the surge of deliveries and ensure that local businesses and jobs aren't annihilated by online commerce? National and state governments are clearing the way for automated trucking. For instance, in Texas "a half-dozen startups are testing in the state, and some are already offering commercial services." A handful of cities, including Paris, New York, and São Paulo, are encouraging night delivery. But today, even most big cities don't know how much freight moves, when, where, or why. If the driverless revolution takes a quick turn down the road of low-cost delivery, cities will be ill-equipped to adapt. But it's still unclear how future changes in shipping will translate into traffic.

Today, trucks make up about 10 to 15 percent of road traffic in the developed world (it's much higher in poor countries due to low private-vehicle ownership). Perhaps we can expect the surge of deliveries caused by next-day, same-day, and instant shipping to deliver a big increase in the number of trucks clogging up our streets. What's worse, the impacts would be felt primarily in residential neighborhoods, where the growth will be greatest, rather than in business districts.

This logic, however, is flawed. In fact, the opposite is more likely—the boom in online shopping will make trucking more efficient and take huge numbers of shoppers off the roads. The reason why is something known in the business as *delivery density*. As more people on your block pile onto Amazon, Jet, or Alibaba to meet their daily needs, last-mile delivery trucks no longer have to race around dropping individual parcels at scattered addresses. They can park at the end of the block while the driver delivers dozens of packages on foot. Meanwhile, studies extrapolating from recent data suggest shoppers might cut the number of store trips they take by as much as half as they shift decisively to delivery. Rather than flooding neighborhoods with traffic, more online shopping could empty them out.

Instant delivery is the wildcard in this forecast. The economies of scale provided by high delivery density may vanish when a separate conveyor trip is needed for every package. According to KPMG, a consultancy, if all of those saved shopping trips—half of today's total—were replaced with *same-hour* deliveries, *one million* parcel-hauling AVs would be needed in the US alone—three times the number of taxis in service nationwide today.

Then there's the rebound effect. Adding to the swarm of conveyors scooting about on their time-sensitive tasks, local businesses will also find many new uses for cheap cargo-carrying AVs. In 2018, Ford began testing the potential for light-duty commercial AVs in Miami. One experiment involved a small dry-cleaning business, which employed a self-driving Ford Transit van to haul dirty clothes from four storefronts to a central laundering plant. Buoyed by the results, which freed up two full-time employees, the company hoped to put the robovans to work in new lines of service, such as delivering low-cost wash-and-fold to nearby college dorms. It's all but impossible to predict how many miles of additional AV travel service innovations like these could add to local streets and sidewalks.

A more difficult dilemma will arise as the biggest online retailers vie for a last-mile monopoly. Amazon is already making its move. A multipronged effort—including the rollout of package lockers, delivery-consolidation incentives like Amazon Day, and the launch of its own shipping company—hopes to finally bring shipping costs in line with revenues. In 2016, the company hemorrhaged an estimated $1 billion a year on shipping for Prime customers alone. By 2019, the company's logistics operation was estimated to carry half its own shipments, up from just 8 percent in 2016. Increasingly, the gambit looks less like cost control and more like a systematic campaign to box others out of local distribution.

The consequences of a last-mile Amazon monopoly would be profoundly negative. Not only could the company deal a death blow to its remaining online competition, blocking them or charging exorbitant fees for local delivery, it could starve out neighborhood businesses in the same way it has undercut its retail partners online. Over the last decade, the company has launched more than 100 consumer brands, largely by emulating the success and then undercutting the prices of stores launched by independents in the company's own Marketplace program. An all-too-likely nightmare scenario, if Amazon established a stranglehold over same-day distribution, would be the extension of these predatory practices toward local merchants. The snowball effect of failing shops, the extraction of small-business revenue that used to get reinvested in local economies, and the loss of local delivery jobs would hit cities hard. As the company aggressively expands into retailing, with its acquisition of

Whole Foods and the launch of the Amazon Go cashless stores, this bad dream doesn't seem so farfetched.

———

THERE MAY BE ANOTHER WAY. Most cities and towns have little power to push back against such world-historic forces. But by leveraging the modest authority they do have, communities might find a lot to gain by putting a human touch back on robot delivery.

If I were mayor, I'd make my stand at the curb. I'd start by banning conveyors from sidewalks. Combine the rise of continuous delivery with the fact that few cities can afford crazy schemes like freight tunnels—and you have all the makings of a colossal curbside logjam and sidewalks swarmed with robots. The world over, cities' authority over these two pieces of turf is undisputed. And it's a change likely to gain easy support. I'd walk my dog in peace, greet my constituents by name, and smile for the cameras while I plotted my next move.

Then I'd go retro, and bring back the porter, those attendants who guard the entrance of buildings and help with baggage. I'd station them at curbs all across town to greet conveyors and mules and hand-deliver packages the last meter to voters' front door. Paying the porter could be a problem. But there are lots of possibilities. Homeowners' associations might foot the bill simply as a cost of maintaining law and order on subdivision streets. Business districts might pool funds from property owners to keep streets clear for tenants and improve the marketability of office towers. With the city attorney by my side, I might even make a valiant attempt to impose a curb-access fee on Amazon and its ilk for every stop, as Paris mayor Anne Hidalgo has, using the funds to establish a municipal porter service of my own. Worst case, I'd issue porter-service franchises to small-time vendors and create some much-needed jobs for veterans, public housing residents, or other needy self-starters.

I'd make the case that not only was my system better for the people, it was better for industry too. By combining the best of machine learning and human intelligence, it would eliminate the perplexing problem of teaching robots to climb stairs, operate elevators, and knock on doors. I'd call up Jeff Bezos to point out that customers prefer a live delivery person, and are reluctant to use his company's package lockers even when offered discounts.

And then the real fun would start. Because once in place, porters would themselves become a platform for grassroots economic development. Beyond keeping sidewalks and curbs clear, porters could insert themselves in the value chain between click and consumer. They'd step in as handymen to provide services to round out the delivery ecosystem—assembling products, putting away groceries, and whisking away packaging. They would make sure that empty boxes ended up in the recycling bin, not the waste stream. And they might help companies come up with new ideas for handling—or better yet, preventing—returns.

I wouldn't stop there, though. Building on my success with porters, I'd take Amazon head on, and launch a municipally owned last-mile network of my own, to ensure that local businesses had access to the goods distribution and freight they needed to thrive. We'd start small, with a few leased mules and conveyors. Our big goal would be closing the loop on the flows of material, energy, and money moving through our community—reweaving circular economies to replace the web of consumption. Porters could even help direct drones to safe landing spots, a high priority as more research confirms early findings that drones, mile for mile, are far more carbon friendly than delivery cars and trucks.

I imagine Amazon would do its best to crush us. But that would be half the fun.

Moneyballing Mobility

In Ayn Rand's paean to individualism, *Atlas Shrugged*, copper industrialist Francisco d'Anconia, the heir to a vast fortune, makes a lengthy speech. "Money is only a tool," he says. "It will take you wherever you wish, but it will not replace you as the driver." Keep these words in mind as the autonomous age unfolds. Because the most difficult dilemmas communities will face will be fiscal in nature. What new sources of revenue will fund badly needed transportation improvements? How can the power of market forces be harnessed while avoiding the curse of regulatory capture, that corrupt state of affairs when industry leaders seduce their supposed government watchdogs to wield state power in the interests of business? These are the concerns that will keep future mayors up at night.

AVs will decimate two essential sources of revenue for many municipalities—parking fees and traffic fines. A survey by the public-administration trade rag *Governing* magazine found that in 2016, the 25 largest American cities collected nearly $5 billion in automobile-related taxes, fees, and fines—an average of $129 per resident. For Toronto, where the annual take of $100 million in auto fees and fines is modest, composing just 1 percent of the total city budget, the threat was nonetheless noteworthy. "The transition to AVs and intelligent transportation systems will require changes in the City's capital and operating budget assumptions and plans," the city's 2015 forecast concluded. In cities like Amsterdam, where fully 27 percent of municipal revenue comes from parking fees, the impact could be catastrophic.

All eyes have turned to the curb, where the 25 largest US cities collect nearly $3 billion annually in parking fees and fines. "This likely represents the lower bound of the potential revenue that could be expected with technology-driven enforcement since a significant share of paid parking is in fact unpaid and a significant share of that is undetected," notes a seminal 2018 OECD report on the topic. But talk of revenue targets is already drowning out mobility concerns. Here's the tagline of a 2018 feature on *Slate*—"The American city is wasting valuable real estate on parked cars." A week later, *Governing* jumped on the bandwagon with its own avaricious headline, "Cashing In on the Curb." The author was Stephen Goldsmith, a Harvard University professor and Sidewalk Labs advisor. Even the sober-minded folks in the room seem to be losing their heads. "It's the most valuable space that a city owns and one of the most underutilized," says Matthew Roe of NACTO, the city transportation officials' club.

I'm all for cities taking their cut of continuous delivery. But it's unsettling that the pitch for curb pricing has so rapidly moved past the mobility benefits and straight to the payout. Curb pricing was initially proposed as a tool for ensuring that in the future, buses for poor people and jitneys for senior citizens will have a place to pull over amid the herds of Ubers and Amazons. But when you hear so-called experts talk about curb pricing today it sounds like a flat-out money grab.

Cities that cash in on the curb through access fees won't be flirting with the financialization of mobility so much as giving it a giant bear hug. The danger in such tactical thinking is that we'll lose sight of the larger battle, and lose control over the public realm. How far are they willing to let the creeping frontier of market logic advance? First roads, then curbs— will sidewalks be the next frontier for congestion pricing? We saw earlier how the ride-hail revolution is leading toward collusion, consolidation, and off-loading of tolls onto consumers. Could this be the future for last-mile delivery too? Congestion pricing at the curb could be manipulated by cash-flush corporations like Amazon to starve out the competition. And it's easy to imagine cash-strapped cities cutting deals with Amazon, trading up-front lump-sum payments for preferential access to residents' front doors.

But far from overvaluing the curb, we're probably underestimating its strategic importance in the future. Perhaps it is time to rethink the curb as a new kind of border between online and local economies, and the parking meter as the twenty-first-century customs house. If mega-retailers like Amazon pose an existential threat to local retail and service economies, curbs will be the quay where their global pipeline of goods is discharged onto local shores. It's the best choke point for cities to push back. Instead of simply charging a fancy, futuristic form of parking fees, perhaps we should tax the value of what's washing up.

10 Pushing Code

The ghosts swarm.
They speak as one
person. Each
loves you. Each
has left something
undone . . .

—Rae Armantrout, "Unbidden"

Writing a book is a peculiar kind of work. There's an initial burst of contagious excitement with your colleagues, loved ones, and friends. But then it is you and your ideas for days, weeks, months, and years on end. You wrestle with pages that are either completely empty or full of words arranged in the wrong order. What surprises you most is the solitude. Your only guides and companions are the works left behind by those who have blazed the trail before you.

Eventually you reemerge, manuscript in hand. Like a normal person doing a normal job, you're looking for some feedback, and some reassurance that you're on the right track. But your first draft always has rough patches, and even the gentlest of critics will quickly find them. And so you return to your desk for rewrites, having taken your first real step developing the acute sense of delayed gratification you'll hone as an author. Now the pressure is on, because while we live in a world of Wikipedia revisions, deepfakes, and redacted tweets—you the author still get but one shot to nail your text perfectly.

Nothing could be more different from the way programmers create code. Their vocation is intensely collaborative, endlessly iterative, and immediately gratifying. Software either works or it doesn't, and the computer lets you know right away.

When I began to work on this book, I already knew a lot about how coders work. But it had been more than 20 years since I'd written any software myself. And there was so much I wanted to experience firsthand in the fast-evolving world of artificial intelligence, to understand its deeper implications for the driverless revolution. And so I decided to dive back in. During the year I spent writing, my days were filled with prose, but my evenings were consumed by code. I didn't build self-driving software, but I did cobble together a mobility app for tracking buses in my neighborhood. I taught myself Python, one of the most widely used computer languages today, and experimented with the open-source software powering AVs— freely shared code libraries that do the basic work of mapping, machine learning, and computer vision.

Along the way, I also learned how mistaken our stereotypes of these two professions are. We mythologize the worldly author sitting in a café drinking wine and smoking cigarettes, discussing important ideas, surrounded by friends. We stigmatize the outcast computer engineer, the "hacker," hidden away in a dark room where he's more comfortable with technology than people. But in reality, the opposite is usually the case. Coders are far more social than writers. They talk in chat rooms and meet up for coffee. They share, constantly, almost compulsively—tips, tricks, hacks, and how-tos. Writers, on the other hand, often have a hoarding problem. They hold on dearly to their notes, story ideas, and sources.

Rough drafts are kept under lock and key. Programmers go to the opposite extreme. They're prolific publishers of works in progress. "Pushing code," they call it.

So even as I wrote as writers do, in solitude, I coded like the coders. The first working prototype of my bus tracker—what they call the "minimum viable product" in Silicon Valley—could display only text and tables. It looked awful, and had all kinds of bugs. I pushed it out anyway and people started to fool around with its features and pore over my code. Following one suggestion, in the next version I added route maps with little markers for current bus locations. A few weeks later, after I'd brushed up on JavaScript, I added scripts to update the positions in real time. Suddenly, buses crawled along the streets on my screen. As more people in the community started using the site, I received more suggestions for improvements. I added letter grades to compute scores for the frequency and reliability of each route, summarizing the hoard of exhaust data I'd stockpiled. Soon, our elected officials began to take notice, nudging the local transit agency to get busy adding service and clearing a couple of particularly bad bottlenecks.

As I experienced this way of working, my understanding of how we'll tame the AV matured. Code isn't simply the raw material of our automated future; it is our most powerful point of intervention. That isn't to say we should dictate the design of self-driving software. But we must set expectations for how it performs.

More important, however, we need to expose the code that controls these menacing machines before they hit the streets. Our current approach is to sit back and hope that our old-fashioned methods of pouring concrete and passing regulations will rein them in. But it's clear that this orchestrating software is sly. Even as it appears to do one thing, it secretly works to other ends. Benign business processes like dispatching taxibots become an exercise in subtle algorithmic discrimination. The simple task of routing rides, multiplied a millionfold each day, allows companies to usurp control from government traffic engineers. That camera we thought was measuring our readiness to take the wheel back from computer control is secretly profiling us for ad pitches. This code is the true battleground, where our assumptions about efficiency, safety, and fairness will be inscribed into algorithms that shape every future movement. This is where the big conflicts over our future will be won or lost.

But even if we recognize the importance of code in shaping the driverless revolution, we need to get much better at writing it. Unlike the materials of the past, which molded the future by laying asphalt and casting steel, code is highly malleable. Unlike a book, it gets revised all the time. That's why programmers have such an astonishing array of processes to work with living code—*commit, revert, rebase, fork*, and *branch*, to name but a few. If you go through the code of my bus-tracker app, you'll see more than 350 milestones that mark where my collaborators and I made small changes.

Similarly, our blueprints for the driverless future must be as malleable as the code AVs run on. They should lay out big, uncompromising goals—but they need to be flexible about how to achieve them. This, however, is not how most local governments build technology today. They plan on 10-year cycles even as the world changes day by day. And so in these final pages, we consider the true code that will shape our future communities—software—and the enormous questions it raises. How much will the limits of today's self-driving software box us in? Conversely, what are the risks that AVs could give rise to AI we might no longer be able to control? And what "code"—ethical, moral, and social—can we, as humans, try to live by as we navigate the choices that the driverless revolution's software shapes for us?

Seeing Like an AV

One of the first books many students of urban design read is Kevin Lynch's *The Image of the City*, published in 1960. Lynch was a contemporary of Jane Jacobs, but lacked her soulful passion for urban social life. Instead, the MIT professor devoted his career to a studied distillation of the physical vocabulary of the city. He tried to decipher how the look and layout of buildings, blocks, streets, and plazas shaped the mental maps people create to navigate cities.

Lynch's big idea was something he called *legibility*, an innate quality of cities that reflected how intuitive their structure was. A printed page, he argued, "if it is legible, can be visually grasped as a related pattern of recognizable symbols." Cities should work the same way, Lynch believed. By simply looking at what's around them, people should be able to easily identify important places, their purpose, and how they connect to one another. *The Image of the City* laid out an elegant and useful vocabulary of five basic, universal components of city form—paths, edges, districts, nodes, and landmarks. Master this intuitive geometric shorthand and you can read any city like a book.

For Lynch, legibility was the make-or-break factor in a city's success or failure. Most cities possessed the basic pieces, but whether or not those parts connected up into memorable patterns—this was what separated great cities from the bad, or merely good, ones. He explored three cities as case studies, illustrating the full spectrum of legibility in urban America. The most legible was his stomping ground of Boston, with its human-scale network of colonial roads, intimate parks and plazas, and historic buildings. The worst was Los Angeles, where Lynch found people disoriented by the lack of features that communicated how its different parts fit together. The third case study, Jersey City, New Jersey, fell somewhere in the middle, alternating between legibility and illegibility from block to block (as it still does, nearly six decades later). Lynch was way ahead of his time, too, highlighting how the components—and their varying legibility among social groups—worked to reinforce the separation of space based on class, ethnicity, and gender.

The legacy of Lynch's work on legibility is everywhere in the way we move about cities today. Urban *wayfinding* (a term he coined) is a large and growing industry. Many cities around the world have invested heavily in coordinated signage schemes that help bridge gaps in the urban "text" and encourage more walking. For instance, Legible London, a massive wayfinding effort launched in 2012, consolidated 32 separate signage systems into a unified network of informative, intuitive sidewalk guideposts. The system has reduced the number of people using the overcrowded Underground for one- and two-stop trips. And when they walk, people burn calories and spend at local shops.

Legibility can also help us make sense of what's to come in the age of automation. For Lynch, there was no legibility without human interpretation. "Like a piece of architecture, the city is a construction in space, but one of vast scale, a thing perceived only in the course of long spans of time," he wrote at the very beginning of *The Image of the City*. He gave us a language for decoding the subliminal messages that great cities send us, and explained how our identities—male and female, black and white, resident and visitor—flavor the conversation between building and self, telling us which way to turn and what to expect when we round the corner.

Yet here we are on the cusp of this driverless revolution—having finally opened our eyes to what our urban surroundings are trying to tell us, only to close them again and outsource our visual perception to machines. Autonomists make a big fuss about how far computer vision has come in such a short time. But considering that few AVs have ventured far from familiar territory, there's good reason to fear this progress has been oversold. Waymo's massive Arizona rollout is a mere 280 miles to the east of California's George Air Force Base, the site of the 2007 DARPA Grand Challenge (often called the "DARPA Urban Challenge" for its simulated city terrain). In all that time, Google's ballyhooed team has merely replicated success in one set of idealized conditions—scant rainfall, ample sunshine, and roads built to modern engineering standards—in a largely identical region. The journey from here to the urban wilds of Dhaka, or even Detroit, will be a long and uncertain one.

"[The city] is seen in all lights and weathers," Lynch continued. So how will a technology that works best on only the brightest and fairest of days lead us into the future?

THE STUBBORN SHORTCOMINGS of computer vision become harder to ignore with each passing day.

Some of the flaws are comical. In Providence, Rhode Island, driverless shuttles got confused by foliage during a 2019 pilot. As it turned out, "the preprogrammed route was mapped when trees were bare. Once the leaves and other foliage emerged, it disrupted the sensors that help the vehicles navigate."

Others are potentially tragic. So much energy has gone into making AVs good at spotting other cars, it turns out they aren't so good at spotting *us*. Ford's AVs, cruising around the college town of Ann Arbor, Michigan, in the summer of 2017, couldn't even detect a crosswalk unless the intersection was equipped with a special signaling device. When an Uber AV became the first to fatally strike a pedestrian, in Tempe, Arizona, in the summer of 2018, a leading legal expert on AVs speculated that the company's software probably "classified [Elaine Herzberg] as something other than a stationary object." And a growing body of evidence indicates that computer-vision algorithms, like the ones employed in Tesla vehicles, may also be better at identifying whites than nonwhites.

AV makers would love to see communities shift gears, from spending money on signs that make streets more legible for people, to rendering streets readable to self-driving machines instead. The marking of pavement to help orient drivers dates to the 1920s, when officials in Detroit repurposed a tennis court line-marking machine to stripe out crosswalks and parking spaces throughout the Motor City. Future streets may feature a funkier look, sporting semacodes instead—those data-rich two-dimensional barcodes. Each of these information-rich glyphs can encode a unique URL, from which the AVs can download an unlimited amount of data detailing rules for that particular block or intersection. As human-driven vehicles disappear, semacodes might even replace conventional street signs altogether. What no one has figured out yet, however, is how cities would pay for all the frequent touch-up work such detailed signs would require.

Another possibility is doing away with street signs and markings entirely, and shifting the display of street rules to cyberspace. That's the thinking at Coord, a spin-off of Alphabet's Sidewalk Labs that's busy building a massive database of cities' street regulations. Beginning in 2017, Coord mappers fanned out across Seattle and San Francisco, driving up and down streets to collect pictures of street signs. Using a custom-built augmented-reality app that picks out traffic indicators from the background and translates their text, the company quickly amassed its own comprehensive catalog of posted regulations. Surprisingly, cities' own records are often incomplete, full of inaccuracies, or scattered across multiple databases. Coord's clever bet is that cities will simply outsource the collation, dissemination, and enforcement of traffic rules instead. It's sure to be a growth market. As the driverless revolution gives rise to more kinds of vehicles, more ways to use them, and new driving behaviors, cities will need more complex regulations.

Coord's platform will also be a great tool for financial engineering to milk revenue from future city streets. Imagine a taxibot entering a city center, destination in hand, looking for a curbside spot to discharge its passenger. The AV, or its dispatching daemon in the cloud, sends a request to Coord's servers. Curbs API, as the company's service is called, writes back with availability and rates for several suitable pull-overs. The taxibot makes a choice, perhaps confirming the exact drop-off location with its passengers, and calls Curbs API one more time to book the spot. The whole transaction is over in a few seconds. Then, as your ride glides to a stop, instead of dropping a dime in an old-fashioned parking meter, it checks itself in. If it hangs around too long after passengers depart, it might get dinged for excessive idling, too. Coord either takes a cut of the curb charge, earns a volume-based license fee from the city, or both. Yet even as the Curbs API vision tantalizes cities with its prescription for total control and free-flowing future revenues, it is an aggressive and risk-filled step forward in the privatization of public space. Seen another way, a lot of what Coord does is scrape cities' own data and sell it back to them. While the company is investing its own resources to compile data cities have ignored, it strikes me as an ethically dubious attempt to make cities dependent upon the company's services in perpetuity.

More city-friendly alternatives to Coord's rent-a-regulation model are being built, too. Washington, DC–based SharedStreets has assembled a registry for street data modeled after OpenStreetMap, a free wiki world map that has served as an alternative to Google's dominance in online mapping for more than 15 years. Unlike OpenStreetMap, however, SharedStreets is more of a closed commons than an open-source repository. As a *data collaborative*, SharedStreets provides a secure, trusted forum for cities, mobility operators, and other stakeholders to share data about streets and what happens on them. The group works with industry partners to develop open curb-data standards that provide an alternative to having Coord as the sole surveyor and archivist for traffic rules. And with funding from Ford Motor Company, SharedStreets has plans to create a framework that mimics much of the functionality of Curbs API. SharedStreets doesn't intend to do the end-to-end hand-holding that Coord is pitching to cities. As founder Kevin Webb explains, "cities need to invest in the mapping in ways that ensure the data remains public information. Crowdsourcing might be part of that but ultimately someone has to put in the time/effort to do the work." But what cities will get in return is ownership and control of this vital virtual infrastructure. Go with Coord, and you'll get a turn-key system. But you'll have to license the data about your own streets to actually use it—and so will every mobility provider, big and small, that wants to roll through your town.

Singularities and Singletons

In 1993, science fiction author Vernor Vinge made a daring prediction. "Within thirty years, we will have the technological means to create superhuman intelligence," he wrote. "Shortly after, the human era will be ended." Computer power was increasing so fast, exponentially so, Vinge argued, that it would soon produce a series of breakthroughs—all-knowing supercomputers, networked brains spanning the internet, interfaces that would blur the boundary between human and machine minds, and biological enhancements to increase intelligence. These, Vinge went on, would give rise to artificial intelligence (AI) far surpassing the power of the human brain. Such a device would be able to continually improve its own design, kicking off a chain reaction of accelerated learning. The Singularity, as Vinge dubbed such an event, would be "an exponential runaway beyond any hope of control," a technological revolution "comparable to the rise of human life on Earth."

This wasn't the first airing for this radical prediction. The mathematician John von Neumann had raised the possibility in the early 1950s. But Vinge's message was timely, concise, and eminently meme-worthy. And while his assertion sounded like the plot for a dystopian novel, many geeks and gurus welcomed the possibility of machines with superhuman intelligence. In the age of the Human Genome Project and climate modeling, the great challenges facing humanity were better seen as informational problems, they argued. All would yield before the awesome number-crunching machines to come.

In recent years, however, the technological Singularity has fallen behind its original schedule. It's behaving quite unlike its cosmological namesake—the point deep inside a black hole where the strength of its gravitational field soars toward infinity. Instead, the Singularity seems to recede more the nearer we draw. Vinge's estimate placed the event sometime between 2005 and 2030. But as he confessed at the time, "AI enthusiasts have been making claims like this for the last thirty years." A recent recalibration by AI pioneer Ray Kurzweil, the technorati's favorite Singularity propagandist, reset the deadline for the digital rapture. "I have set the date 2045 for the 'Singularity' which is when we will multiply our effective intelligence a billion fold by merging with the intelligence we have created."

More revisions to the timetable may be in store. Reports of an imminent "AI winter" are probably overblown—almost no one in the field expects a return to those repeated periods of disillusionment with the technology that occurred several times from the early 1970s through the late 1990s. But there's a gathering unease about the pace of AI development.

During the AI winters, researchers cooked up euphemistic names like "machine learning" to disguise the true nature of their AI work from funders. But today's anxieties stem from a fuller reckoning of techniques that are widely appreciated in the AI community but whose capabilities are poorly understood beyond. Much of the angst revolves around deep learning, which has been weaponized in rhetoric far beyond its technical potential. As Rodney Brooks, the former director of MIT's Computer Science and Artificial Intelligence Laboratory, points out, the phrase is often mistakenly used to suggest that there is "a deep level of understanding that a 'deep learning' algorithm has when it learns something. In fact the learning is very shallow" in comparison to genuine human learning. And while the technique's commercialization has produced enormous economic value in medical diagnostics, natural language translation, and logistics—among other domains—its severe limitations in complex real-world settings are becoming clearer. "Contemporary neural networks do well on challenges that remain close to their core training data," writes New York University computer scientist Gary Marcus in an exhaustive 2018 critique of deep learning, "but start to break down on cases further out in the periphery."

Nowhere are the limits of deep learning becoming clearer than in the development of self-driving vehicles. The most catastrophic failures involving automated driving so far have all occurred around so-called edge cases, those unexpected events where data to train deep-learning models was insufficient or simply nonexistent—a pedestrian walking a bicycle across a darkened street midblock (Tempe, Arizona, March 18, 2018); a white truck trailer occluded against a brightly lit sky (Williston,

Florida, May 7, 2016); an unusual set of road-surface markings at a highway off-ramp (Mountain View, California, March 23, 2018).

In these, and in many other wrecks to come, it is humans who have paid the price for deep learning's shortcomings. But the collateral damage to AI development could be substantial. "If . . . driverless cars should also disappoint, relative to their early hype, by proving unsafe when rolled out at scale, or simply not achieving full autonomy after many promises," Marcus concludes, "the whole field of AI could be in for a sharp downturn, both in popularity and funding."

There may be one silver lining for humans in a self-driving blowout—preserving the pecking order at work. When Oxford economist Carl Benedikt Frey tallied his predictions of the jobs most likely to be lost to automation, there was one that ranked close to last—aircraft cargo-handling supervisor. Perhaps an indefinite postponement of the Singularity will provide plenty of jobs for people whose sole task is to to watch over robots of more limited wits.

AS CONTINGENT AS it may be, you should dismiss the Singularity at your own risk. Because if it *does* come to pass, and the AI it gives rise to is a malevolent one, the consequences are so frightening it is difficult to imagine. One philosopher who has pondered this is Nick Bostrom, whose remarkable 2014 treatise *Superintelligence* imagines that in the near future, human efforts may indeed create an artificial intelligence that can rapidly improve its own capabilities. As the newly sentient entity evolves, in ever-accelerating cycles of self-improvement, its capabilities will grow with stunning speed. From our biological perspective of time, it might simply appear to pop into existence without warning. And it would have godlike cognitive abilities if possible.

If you find the birth of an ultrapowerful AI startling, what comes next will shock you. Tempted as we might be to tap this gizmo's talents for our own aims, by this point it would already be too late to control our creation. As Bostrom explains, assuming we're even alert to its existence—which it might go to great lengths to conceal—it would be too late for us to change the software of a superintelligence without its consent. Even if we could bypass its safeguards, its improvements to its own code might have already outstripped our relatively unsophisticated knowledge. And so it would be free to pursue whatever goals it had inherited from its original programming, and those it had since cooked up on its own. If such a device were to achieve a "decisive strategic advantage—that is, a level of technological and other advantages sufficient to enable it to achieve complete world domination," it could seize control and form a *singleton*—Bostrom's ominous term for "a world order in which there is at the global level a single decision-making agency."

It's easy to imagine some of the nightmare scenarios. A rogue AI emerging from deep within the Amazon cloud, subjugating humanity to its aspirations of infinite supply-chain efficiency. Some self-driving fleet, playing out the sad script of that 1935 sci-fi story we saw in Chapter 4, "The Living Machine," and running rogue across the land. And the scariest part of Bostrom's postulates? It's hard to prove that this hasn't already happened and that we aren't living under a singleton already, like Keanu Reeves in *The Matrix*.

What makes the prospect of a singleton so disconcerting, however, is that it's not merely possible, Bostrom argues—it's inevitable unless we take precautions to prevent it. Runaway superintelligence accidents, it seems, are more like nuclear accidents than rogue meteor strikes. We need to build in safety measures. And those may still backfire, as they did at Three Mile Island and Chernobyl.

The bulk of *Superintelligence* isn't so much about scaring the shit out of us (which it does very well) as it is about starting a serious and sober discussion about what to actually do about the threat (which it has also done). Bostrom's vision alarmed enough influential people—Bill Gates and Stephen Hawking, among others—to trigger a soul-searching about the future of AI that is still unfolding. Gates and Hawking are among the backers of Bostrom's not-so-humbly-named Future of Humanity Institute at Oxford, which has raised more than $25 million since 2015.

The driverless revolution hardly needs superintelligence. Self-driving cars and trucks will make do with far less robust AI. But automated mobility will certainly provide a rich spawning ground for a singleton to emerge. Supercomputer-powered AVs, deployed in highly coordinated fleets, will be the most massive apparatus for machine learning ever assembled. Add in the awesome complexity of software tuned to manage planetary flows of people and goods, and the unpredictable interactions among billions of instances of running code, and the probability we're incubating something sentient quickly grows. A greater risk, however, may lie within the dark web of mobility markets, and the computational colossi of tomorrow's traction monopolies like SoftBank, Amazon, and Alphabet. Bostrom conjectures that a potential path for a superintelligence to attain power is by "subtly manipulating financial markets." He compares an aspiring singleton's inevitable ambitions to tap terrestrial resources and take control of civil infrastructure to the Nazis' dash for the oil fields of Romania. Once a would-be digital dictator sees the chance to take a cut of every AV movement on the planet, it could make its move with a speed and force that will make today's would-be traction monopolies look tame by comparison.

Big Mobility

Throughout most of human history, the movement of populations left its mark on the earth and in the heavens. Caravans kicked up dust clouds visible for miles, and wore tracks in the dirt. But when we paved over those rutted trails, the toll of our travel on land and sky vanished.

Now, as the driverless revolution begins, we are building dynamos of code and data to propel even more traffic through these asphalt pipelines. The three big stories of this book point toward a common potential—a vast explosion of ideas and inventions for travel. There will be more kinds of vehicles, more services to meet our mobility needs, and a vast new capacity to finance it all through markets for movement. Left unchecked, these advances could unleash an uncontrolled expansion in the movement of people and goods. Unless we change our ways there will be no containing the colossal, catastrophic spillovers that result. If we are wise, however, these same shifts could serve as tools to deal with the pressing threat of climate change. The ghost road beckons, then, with promises and perils.

Throughout this book, we've seen many good policies and best practices that will help prepare us. There are already a *lot* of good ideas about what the AV industry and government must do, and more are sure to come. But it isn't enough to let institutions take the lead. As individuals, as families, and as communities we face a set of choices about how we travel, the products and services we buy, and the activities we engage in every day. These choices will shape mobility markets and may matter more than any efforts our leaders can organize in the next few years.

But how are we to live? In these final pages, I propose a set of principles for a more aspirational, thoughtful living that I call *big mobility*. Think of it as a code for humans in the driverless revolution that distills what we know about the future to make better decisions today. Big mobility makes three assumptions about the future.

First, big mobility is technology-enabled. It embraces automation as a means to reduce the cost, carbon footprint, and danger of vehicle transportation. We shouldn't rush to place restrictions on AVs, but we also mustn't shy away from exercising strong oversight on their development and use. And we must search for ways to exploit their efficiency and flexibility, without compromising our values.

Second, big mobility is travel-intensive. Nobody should live in a community where bad design or bad policy forces them to travel unnecessarily. That's why progressive approaches to urban planning emphasize *access* over *mobility*. The best community is one where there's no need to travel at all—everything is already close to everything else. While this is a noble aspiration, rebuilding entire nations and cities in this way takes too long. It's time to face up to the fact that while we have largely failed to find ways to build for access at scale, our ability to invent cheap, flexible, and sustainable mobility technologies like rovers, conveyors, taxibots, and software trains is remarkable. We must continue to push for better community design, moving away from sprawl and private vehicles toward walking and transit. But we also need to accept the urgency of the adaptation challenge at hand. That means we must exploit the cheap, flexible mobility AVs provide to expand into new towns built on new designs, rather than wait years for contentious changes in land use within existing communities to be agreed on.

Third, big mobility is self-indemnifying. That is to say, it actively measures and mitigates the externalities of all that travel, revealing harms like congestion and carbon emissions, in order to influence our behavior. It elevates transparency and accountability to the same level of importance as traditional indicators of transportation system performance like frequency, on-time arrival, cost, and reliability. This means big mobility puts constraints on mobility markets in order to hedge against the future risks—environmental, social, and financial—that they create.

The goal of the six principles that follow is straightforward—to steer you, ever so slightly, down a more benign branch of the ghost road. Use them as a filter, to help make sense of future choices about AVs in your life and your community, whether you are clicking on an app or at a voting booth. With luck, they will help you achieve that ancient dream, the effortless fulfillment of our innate human desire for propinquity, but do so while respecting the extreme constraints our overstressed planet now places on us.

First, demand equal access. When Waymo started routing self-driving taxis through metro Phoenix in 2018, homeowners in a handful of neighborhoods responded with protests and threats of violence against vehicles and safety drivers. During one incident involving an intoxicated local staring down a Waymo van, a company test driver told police "that she would notify Waymo to stop routing vehicles to that area." Did Waymo actually delete the subdivision from its service map? Even the possibility of such a move raises disturbing concerns for the future.

By making travel cheaper, easier, and more spontaneous, AVs could generate enormous benefits for mobility-challenged groups such as the elderly, people who don't own cars but live far from mass transit, and teenagers. But if AVs are to work for everybody, everybody must be able to use them. And choosing whether or not to serve some community isn't the only way companies will write subtle exclusions into AV services. Prices will provide powerful, but often pernicious, signals about who and what can move where, when, and how. Categories, encoded by machine-learning algorithms themselves trained on biased data, will perpetuate harmful stereotypes below the radar.

Much of the might to prevent these abuses rests with regulators. When companies can cherry-pick to whom and where they provide service, without scrutiny, they are making mobility policy without public consultation. But you can vote with your wallet, by taking transit when possible, and by boycotting AV companies that engage in redlining and algorithmic discrimination.

Second, fight separation. You'd think that the threat of physical separation would pale in comparison to all the algorithmic slicing and dicing to come. But this idea is the zombie solution of traffic engineering—an idea that's been killed and buried many times, yet keeps coming back to life to haunt us anew. There's almost no limit to how dark the designs could get. Consider a 2016 patent filing, unearthed by AI ethics researchers that revealed Amazon's early thinking on the proper place of workers and AVs. As the *Seattle Times* reported, "illustrations that accompany the patent . . . show a cage-like enclosure around a small work space sitting atop the kind of robotic trolleys that now drive racks of shelves around

Amazon warehouses." It's as jaw-dropping a metaphor for an undesirable, yet possibly unavoidable, urban future as you're likely to find.

Thankfully, Amazon shelved its human-cage plans. Instead, many of the company's fulfillment-center workers now don a less repressive protective device to ward off warehouse droids, called the Robotic Tech Vest. By signaling a human presence to nearby machines, this ghost road armor allows more than 125,000 fulfillment-center workers to commingle with some 100,000 warehouse droids, who quickly map out human habitats. "In the past, associates would mark out the grid of cells where they would be working in order to enable the robotic traffic planner to smartly route around that region," explains an Amazon robotics executive. "The vest allows the robots to . . . detect the human from farther away and smartly update [their] travel plan to steer clear without the need for the associate to explicitly mark out those zones."

Amazon's protective gear is better than the android apartheid the cage contraption would create, but it still puts us on a slippery slope. If such schemes spread, we may soon all find ourselves donning doodads that mark us as living liabilities. City planners' gospel for the driverless age, the *Blueprint for Autonomous Urbanism*, instead takes a human-beings-first stand on the matter—"pedestrians [should be] connected not detected." Unlike in Amazon's warehouses, "people walking and biking should not be required to carry sensors or signals to stay safe" and "vehicles should be able to detect and yield to pedestrians in all conditions, and retain full responsibility for not injuring people."

So what are we to do? The thought of deep-learning AVs encroaching ever more on our personal space is deeply disturbing. But unless we can hold self-driving machines to a high standard, demanding they deal with us on our terms, and accept them only when they prove capable—the alternative could be far worse. Instead of taming the AV by teaching it ourselves, we'll banish it to a foreboding place where machines dominate and people—behind the wheel, on bike, or on foot—are fearful. Not only will this manifestation of the ghost road put much of our world off-limits, like the automobile did during the twentieth century, it will increase the economic and social costs of automation. Whatever infrastructure is required—separate road networks, machine-legible signage, new wireless beacons, and more—will be financed by taxpayers. And we'll all pay the larger costs of more separation.

Third, don't get demobilized. Automation could turn us all into the worst kind of couch potatoes. In rich countries today, few people meet recommended daily exercise guidelines. The driverless revolution could trigger an ever-further-downward spiral. As local services ditch storefronts, walking or biking to the hairdresser or dry cleaner will be replaced by taxibots bringing services to your front door. I won't deny it—there's something almost magical about same-day delivery. You can now get a vast universe of goods delivered to your home in far less time than you can get your act together to go to a store and buy something. But such conspicuous consumption always takes a toll. This time it will be on our waistline, our mental health, and our main streets.

The hard work here will be making sure that some of the dividend of the driverless revolution is reinvested in things that keep us active. If microsprawl replaces pedal bikes with personal EVs, we need to provide new centers for open space and outdoor exercise. If main streets hollow out as retailers ghost into the fulfillment zone, we must create new hubs for the casual civic life that keeps us connected. As work moves online, we'll need to generate opportunities for meaningful livelihoods that keep us from becoming isolated in our own homes, away from the social safety nets that, science tells us, keep us healthy and happy.

To fight the urge to sit back, change your own behavior. Don't use ride-hail for any trip under a mile on foot, or up to five by bike or bus. Try installing a shop-blocking browser extension like Icebox, which will quarantine your online impulse buys for up to 30 days—long enough to consider shopping locally instead. Look nearby for goods and services, pricing in the negative impacts of speedy delivery, and get off your ass and out into the world instead.

Fourth, use *vehicles*—don't own them. There's far too much hope being placed on ride-sharing as a panacea for the excesses of the automobile age. But not only are shared rides even more heavily subsidized by ride-hail operators today (they lose more money on shared rides, which are seen primarily as a customer-acquisition and brand-building strategy), sharing back seats with strangers has plateaued far below levels needed to bring about the transformation desired by car-lite communards. It is but hopeful speculation that the fare cuts brought on by automation will change this fact.

Automation can still help us move away from single-passenger cars, however. The driverless revolution will deliver an endless variety of smaller vehicles and specialized services. That's the payoff from our devil's bargain with financialization. So rather than forcing people to share sedans, we should structure mobility markets that allow companies to discover new ways to motivate people. Lower fares aren't the only carrot that can entice people to share rides. The financialization of mobility carries many risks, but it may be highly effective at targeting new kinds of incentives that actually do increase vehicle occupancy, if we let it.

Your role in this chapter of the driverless revolution starts, then, with selling your car and starting a subscription. Today, Uber and Lyft are just beginning to experiment with such offerings. A steep learning curve lies ahead as they figure out how to price so many possible options through a single app. Will your tastes tend to a specific mode—taxis, bikes, scooters, or transit—or favor specialized services like kid-friendly cabs or senior shuttles? Packaging and cross-promotion will play a big role, too. We'll see mobility services bundled with other products, like housing. And as internet giants colonize our communities, "triple plays" that offer rides, groceries, and a multimedia cloud for one monthly fee—an entire iLife, if you will—could become commonplace.

Of course you can still pay as you go. And a handful of cretins may still prefer to own their own cars. But their numbers will shrink, and the premium they'll pay will grow as congestion tolls and carbon taxes add up. The rest of us won't notice as those pesky fees are simply folded into our monthly bill.

Fifth, find your own uses for self-driving things. People are always portrayed as passive consumers in the driverless future. This needs to stop. The way we'll use this technology is not something that anyone can, or should, try to predict. Real people repurpose new technology in ways the inventors never imagined. This timeless urge to tinker will be a primary force shaping self-driving vehicles and what we do with them. As cyberpunk author William Gibson wrote in 1982, "The street finds its own uses for things."

Car enthusiasts, for instance, have always cracked open engines to push the limits of manufacturers' designs—from the earliest days in Detroit to the 1950s hot-rod gangs in Los Angeles. But as I've argued throughout this book, the programmability of AVs is their most important and novel advantage. These machines are, from top to bottom, utterly malleable from one instant to the next. The potential for innovation by end users is almost infinite.

Obviously we don't want people overriding safety controls and putting others at risk on public roads. So it isn't yet clear how companies will open up AVs to tinkerers, or how hackers might jailbreak driverless cars' code to do so themselves—or whether such hacking would even be legal. An array of tools, from the infamous end-user license agreements (EULAs) to subscriber contracts, will limit what is permitted under civil law. A wide array of road-safety and cybersecurity laws will soon add criminal penalties for the riskiest kinds of tinkering, too.

But there will be spaces for experiments and we should all give them a go. Just as with a smartphone today, Google or Apple may provide the underlying operating system that senses the road and spins the wheels.

But when we're on the move in the driverless future, we'll turn to apps for all the task-specific tweaks we need to inform and organize, enhance and optimize, and simply liven up our travels.

Sixth, let down your guard. And then look up. Throughout this book, I've ignored drones. The omission is deliberate. Practically speaking, drones are such a complex and speculative topic, it would be difficult to do them justice while also surveying the vast terrain covered by developments in terrestrial AVs. And while drones are already here, in surprisingly large numbers—more than one million are registered in the US alone—they simply aren't going to factor into the urban equation anytime soon.

Unmanned aerial vehicles (UAVs) are actually easier to engineer than self-driving cars. Once you figure out the physics of staying aloft, there's simply less to hit up there, even in cities. Tall buildings can be mapped far more precisely, and updated more regularly, than the chaos of the streets below.

The real obstacles for drones in populated areas are political and bureaucratic—they are the ultimate NIMBY nuisance, and things that fly are, for good reason, wrapped in red tape. Everyone wants the convenience of drone delivery, but no one wants to have drones flying down their street all day, and they definitely do not want a drone depot at the end of the block. It's hard to imagine a use of public airspace with more unpleasant impacts and such superficial benefits as swarms of buzzing drones dropping off burritos and gift baskets—typical uses for urban drones. Moreover, aviation regulators are among the most risk averse in government. Given that drones will operate above streets filled with people, we can expect all the usual concerns about airworthiness, flight-path obstructions, and emergency landing zones to be amplified manyfold.

If we let down our guard a bit, however, we might find ourselves more willing to welcome at least one special class of municipal drones, a new breed of public-safety auxiliaries I call *guardian angels*. If you have ever experienced a major urban disaster, you'll immediately see the appeal. Tasked with a singular mission as the stresses of climate change mount—

improving urban resilience—during peacetime, guardian angels would maintain the walls. They'd inspect man-made defenses like dikes and dams; monitor natural barriers like dunes, wetlands, and oyster beds; keep escape routes along highways and bridges secure; and even make preventive repairs and replant trees and grass. In a crisis, guardian angels' mission would change to helping humans directly: guiding evacuations, locating people in distress, mapping damage, delivering relief supplies and medicines, and providing communications relays for responders and residents. Guardian angels might be dedicated droids, or simply delivery drones deputized on the spot. Others might do double duty, snapping aerial photos for government maps as they go about their daily business, perhaps—as John Edgar of Stae, an urban-data startup, suggests—as payment-in-kind for use of city airspace and air-traffic-control services.

Urban drones will come in good time. Compelling commercial and public-safety applications like guardian angels will create growing pressure to clear urban skies for low-altitude UAV operations. Building owners will lease their rooftops as droneports to speed rollout. Architects will pierce the exterior walls of buildings to provide portals for new ductwork and drone chutes that provide easy access to interior rooms. And urban planners will find space that might do double duty as crash-landing zones, such as the medians of busy streets. But as long as drones take to arrive, these adaptations will take even longer.

For now, the best you can do is plant a tree. David King, the transportation scholar, wonders if "drone noise becomes [the] thing that pushes cities to really invest in street trees—because the canopy provides a natural sound barrier." So if you bury a seed today, by the time drones arrive on your block, you should be in the shade.

A Fork in the Road

It's said there are 100 million lines of computer code in the average car rolling off the assembly line today—a figure that is growing exponentially as automation takes over. Billions more shape the cloud, keeping tabs on vehicle fleets and on the endless machinations of markets that balance their supply with our demands. It is a text we will never finish. With every line, the risks of catastrophic, cascading failures multiply.

The ghost road will be paved with code, too. But here, code's malleability works in our favor. Everything we do today is provisional. It's all subject to change. We can design a different vision for the driverless revolution to come.

In the same spirit, this book is also something of a provisional work, subject to extensive and frequent revision as the future actually arrives. But the three big stories—the bewildering variety of automated mobility, the transformative power of near-zero-cost freight, and the stealthy risks of financialization—are, I believe, here to stay. These are the forces that will decisively shape every step of the driverless revolution to come.

We've come as far as we can together. It is time for you, like a young programmer, to enter that future world of strange new code on your own. As you do, you may soon find yourself on GitHub. It's a site, kind of like Wikipedia, where people all over the world collaborate on programming projects. Spend some time there, and one day you'll find yourself staring at someone else's work when you suddenly realize that *you can do better*. Your first move will be to create a *branch*, a temporary copy of the original work that lets you build and test your enhancement before weaving it back into the main trunk. The process of working on branches is kind of like writing some revisions at the next desk over, and handing your edits in at the end of a day. You're tinkering, inching the larger collective effort forward.

But the time will come when you see a need to move some project in a whole new direction, break with the current consensus, and challenge the status quo. For that you'll need a *fork*. Like the proverbial fork in the road, GitHub's scheme takes you down a new path—it's more like walking out the front door with a hard drive of the whole venture. It isn't clear if you'll ever come back.

Forks are as risky in coding as they are on the open road. Suddenly, you're on your own. As you work in solitude, your inkling of an idea may die on the vine before it attracts new followers. But it may also point the way toward far greater gains than could be achieved the original way. This is one of the oldest tensions that innovators in any society face. As the poet Robert Frost wrote:

Two roads diverged in a wood, and I—
I took the one less traveled by,
And that has made all the difference.

Take a few steps toward your dream for the driverless revolution—whether it's by writing code, changing your consumption, ditching your car, or something else entirely—and you may quickly find you've blazed a trail many others will follow.

So take a good look around now, and gather up your things. I've had my say. Take this copy and go now and start writing your own version of what's to come.

Go ahead. Fork me.

Afterword

Nothing has spread socialistic feeling in this country more than the use of automobiles. To the countryman they are a picture of arrogance of wealth with all its independence and carelessness.

—Woodrow Wilson (1906), president of Princeton University

In the early 1920s, the writer Thomas Wolfe left his childhood home of Asheville, North Carolina, to move to New York City and make it as a playwright on Broadway. He didn't find success on the Great White Way, but turned his talents instead to all but single-handedly invent the uniquely American genre of autobiographical fiction. In his final novel, *You Can't Go Home Again*, published posthumously in 1940, Wolfe lamented the transformation that life in the big city and the financial pillaging of the countryside had wrought on both his self and society at large. "He saw now that you can't go home again—not ever," Wolfe wrote. "There was no road back."

I didn't set out to write another book about cities. Like Wolfe, I'm a small-town kid who chose the metropolis as home. For me, though, there is a road back—a 150-mile stretch of the Garden State Parkway that links the sprawl of New York City's suburbs with the long, blue Atlantic shore. It's a leisurely drive, twisting through the eerie hollows of the Pinelands National Reserve, the haunting ground of the legendary Jersey Devil. I often make the run late on a Friday night, to avoid the gamblers and sun-seekers headed to Atlantic City and the beaches beyond. Finally, at the end of the land, under a sky full of stars, I pass over one last bridge and onto the barrier island. When I arrive there, I open the window and let the sound of the sea surround me.

The city is my life's work, but I feel the pull of this faraway place more each day. Nowadays, when I make this trip, my mind wanders through the possibilities of extreme commutes powered by automation. Could I live in the country and work in the city? I could be up before dawn and into the car, drifting off as I pass by the docks—as the fishermen set off on their own sleepy, self-driving commute to points offshore. The daily grind would take its toll. But it would be a small price to pay for a life in this jewel of a coastal town.

The daydream doesn't last long. Another thought process has been running in the background of my brain, tallying the cost of all those extra miles. Do it daily, and we're talking 70,000 miles a year—more than four times the distance the average American drives each year. I've let myself become intoxicated by this technology's possibilities. And as I imagine the rising cloud of carbon and the trail of congestion I'd leave behind, I find the willpower to chase these sinful thoughts away.

Still, the temptations of self-driving sprawl have a way of creeping back in. I walk along the beach and ponder the dilemma of a push-button supercommute. My head says this way of living will be unsustainable. My heart has another story. Then it hits me that soon, I won't be alone in facing this future dilemma. And if *I'm* already this conflicted, what happens when millions of Americans face the same choice—with far fewer reservations to hold them back?

IN AMERICA, the divide between center and periphery now maps to a deep factional schism. Our geographic self-sorting into red and blue has left the country unsure of its future, fearful of its present, and schizophrenic about its past. And as we imagine the age of AVs, we tend to extrapolate a technologically perfected version of what's right in front of us today. One side sees car-lite communes, the other a promised land of self-driving suburbs. But the makers of self-driving technology don't really care which path we take. So long as the market is maximized, they're happy to bring us together or fling us apart. The choice is ours.

Today, no other rich country depends so heavily on automobiles—three of every four Americans drive to work alone each weekday. Let's assume, for argument's sake, that the three big stories of this book are on the mark. Specialization, continuous delivery, and mobility's financialization will define the driverless future. Together they'll be a powerful centralizing force, pulling population and economic activity back toward the core of metropolitan areas. But will their combined pull be enough to overcome the dispersal caused by decades of policies that distort markets for land, housing, transportation, energy, and water? Perhaps. But we can't leave such a critical choice to chance.

No other developed country will grow as much as the US in the coming decades. Assuming current levels of immigration are maintained, the US is expected to add more than 100 million people over the next 35 years. That means we need to build about 35 million new homes. This makes for an easy forecast to remember—35 million homes, in 35 years, for a 35 percent expansion of the population.

Finding room for all these residences will not be easy. Infill housing in existing built-up areas is one appealing option, and it could benefit from the changes we've explored here. Imagine towers rising by the thousands on freshly fracked downtown parking lots, and hipsters occupying the empty shopping strips of the fulfillment zone. One study that inventoried California's infill capacity found room for about four million units. Scale that scope up nationwide, and—voilà!—we might barely squeeze everyone in without breaking new ground. But we'd have to forgo parks, since achieving this kind of infill density would leave no room for open space. And forcing it down the throats of neighborhoods would require an authoritarian exercise of state and federal power to crush NIMBY opposition. What's far more likely is that many—if not most—of these 35 million future American homes will be built farther out, on open land. That would make the metropolitan expansion of the driverless revolution as big as any we've experienced before—the post-war suburban boom, or the Clinton-Bush exurban bubble that collapsed in 2007. And we won't just build new subdivisions. We'll need entire new towns—hundreds or even thousands of them.

Our challenge, then, is to use this metropolitan metastasis as an opportunity to reconcile our conflicting visions of self-driving suburbs and car-lite communes. AVs are an important tool, but American communities are already restructuring in ways that create opportunity to rethink the marriage of land use and transportation. Over eight million Americans now work from home, for instance—more than 5 percent of the workforce, and the share is rising fast. Instead of designing these new towns for people with supercommutes of more than 90 minutes each way, could we make them attractive to people with *no commute*? Rather than forcing people to choose between high density and auto dependence, could we harness microsprawl to create car-free neighborhoods with cheaper housing and more open space—as well as all the low-carbon conveniences of cities, like education, health care, and entertainment? Instead of simply speeding the delivery of Amazon orders, could municipal conveyors be the glue that makes circular systems for local manufacturing, retail, and services viable? And for the rare occasion when one *needs* to leave, could AV-powered transit link these satellite towns with cheap, convenient connections? (For what it's worth, all of the schemes we develop for new towns will shape how we do infill, too, especially on larger sites like mothballed shopping malls.)

I'm optimistic that Americans will seize this opportunity. After all, we invented the mass-produced suburb. This is a chance to carry forward what was good about that way of living and leave behind its unsustainable assumptions. But there's no time to lose. Climate change is pressing down, limiting our choices more each day. And the US won't be the only place building dream towns in the driverless revolution—China, Canada, and many other nations will do so too. If we're to lead the way, we'll need to act now to adopt favorable policies for housing, energy, and transportation; mobilize national investments; keep progress in artificial intelligence work moving forward; and deploy the roadside infrastructure necessary to ensure that machines can move safely among us instead of rolling over us.

If we succeed, we may once again become a model for the world. And perhaps, now and then, I'll be able to commute from the beach in my sun-powered, self-driving car with a clear conscience.

What could be more American than that?

Acknowledgments

Many people helped bring this book to life. Brendan Curry at W. W. Norton and Zöe Pagnamenta and her team at the Pagnamenta Agency once again showed their mettle, embracing this project with enthusiasm, helping me reach an ever-growing audience across the world. Sarah Johnson lent an exceptionally clear eye to the copyediting process. Bee Holekamp expertly shepherded the manuscript through the production process. Bryan Boyer, Ritchie Yao, and the team at Dash Marshall helped shape many of the design ideas in this book, and brought them to life in illustration. Melissa De La Cruz provided tireless assistance with research and manuscript preparation.

A number of organizations supported my research. Jim Anderson and Stacey Gillett at Bloomberg Philanthropies; Jennifer Bradley at the Aspen Institute's Center for Urban Innovation; and Benjamin de la Peña, during his time at the Rockefeller Foundation. I'm lucky to have such committed partners who are also willing to engage in the work of thinking these questions through together. Through their generosity I was able to convene a stellar group of people for a workshop in April 2017 that generated many of the future scenarios of automated vehicles found throughout the book—Shaun Abrahamson, Varun Adibhatla, Francesca Birks, Garry Golden, Eric Goldwyn, Greg Lindsay, and Chelsea Mauldin.

Many others infected my thinking along the way. Among them,

David King and Kevin Webb were frequent correspondents, generously offering ideas and insights. Sarah Kaufman's work has been a source of inspiration on the importance of gender in how we think about transportation futures.

Finally, all the folks at Paragraph Workspace for Writers in New York City provided a sanctuary where, day after day, it was possible to find refuge from the beeps and buzzes and write about this crowded, messy world of technology we live in.

Notes

Preface

xiv **private-sector funding ... surged tenfold**: Aparna Narayanan, "Self-Driving Cars Run into Reality—and Are Further Away Than You Think," *Investors Business Daily*, May 24, 2019, https://www.investors.com/news/self-driving-cars-hit-delays-driverless-cars-timeline/.

xv **perfecting the technology will be trickier**: Jeffrey Rothfeder, "For Years, Automakers Wildly Overpromised on Self-Driving Cars and Electric Vehicles—What Now?" *Fast Company*, July 10, 2019, https://www.fastcompany.com/90374083/for-years-automakers-wildly-overpromised-on-self-driving-cars-and-electric-vehicles-what-now.

xv **GM's cruise ... software sometimes failed**: Cohen Coberly, "GM's Self-Driving Car Fails to See Pedestrians and Detect 'Phantom' Bicycles, Report Claims," *Techspot*, October 24, 2018, https://www.techspot.com/news/77083-gm-self-driving-car-division-facing-technical-challenges.html.

1. Fables of the Revolution

3 **"We tend to overestimate"**: Susan Ratcliffe, ed. "Roy Amara 1925–2007, American futurologist," *Oxford Essential Quotations*, 4th Edition (New York: Oxford University Press, 2016), doi:10.1093/acref/9780191826719.001.0001.

4 **the introduction of the steering wheel**: Dan Albert, *Are We There Yet? The American Automobile Past, Present, and Driverless* (New York: W. W. Norton, 2019), 19.

4 **dutiful animals would simply keep following the road**: Markus Maurer et

al., *Autonomous Driving: Technical, Legal, and Social Aspects* (Berlin, Germany: Springer, 2016), 2.

5 **"as if a phantom hand were at the wheel"**: "Radio-Driven Auto Runs Down Escort," *New York Times*, July 8, 1925.

5 **a death rate 18 times higher than today**: "Car Crash Deaths and Rates, Historical Fatality Trends," National Safety Council: Injury Facts, accessed January 21, 2019.

5 **a "traffic control tower"**: Albert, *Are We There Yet*, 248.

5 **guide wires embedded in the road surface**: Evan Ackerman, "Self-Driving Cars Were Just around the Corner—in 1960," *IEEE Spectrum*, August 31, 2016, https://spectrum.ieee.org/tech-history/heroic-failures/selfdriving-cars-were -just-around-the-cornerin-1960.

5 **price tag for guided-vehicle highways**: Author's calculation. In 1969 Ohio State University researchers Robert Fenton and Carl Olson projected the cost of upgrading highways for guided vehicles at between $20,000 and $200,000 per lane-mile ($133,000 to $1,330,000 in 2016 dollars). Robert E. Fenton and Karl W. Olson, "The Electronic Highway," *IEEE Spectrum*, July 1969, 60–67. According to the US Department of Transportation Federal Highway Administration, in 2006 the Interstate Highway System spanned some 214,812 lane-miles, constructed at a total cost of approximately $400 billion (in 2016 dollars), or approximately $465,500 per lane mile.

6 **early droids followed white lines**: Taylor Kubota, "Stanford's Robotics Legacy," *Stanford News*, January 16, 2019, https://news.stanford.edu/2019/01/16/ stanfords-robotics-legacy/.

6 **first AVs used two video cameras**: Marc Weber, "Where To? A History of Autonomous Vehicles," *CHM Blog*, Computer History Museum, May 8, 2014, https://www.computerhistory.org/atchm/where-to-a-history-of-auton omous-vehicles/.

6 **retrofitted a Mercedes-Benz van with self-driving gadgets**: Janosch Delcker, "The Man Who Invented the Self-Driving Car (in 1986)," *Politico*, July 19, 2018, https://www.politico.eu/article/delf-driving-car-born-1986-ernst -dickmanns-mercedes/.

7 **Stanford University's winning vehicle**: Sebastian Thrun et al., "Stanley: The Robot That Won the DARPA Grand Challenge," *Journal of Field Robotics* 23, no. 9 (2006): 661–92, http://isl.ecst.csuchico.edu/DOCS/darpa2005/DARPA%20 2005%20Stanley.pdf.

7 **Silicon Valley moved forward**: Lawrence D. Burns and Christopher Shulgan,

Autonomy: The Quest to Build the Driverless Car—and How It Will Reshape Our World (New York: Ecco, 2018), 137–57.

7 **Larry Page's lifelong interest in AVs**: Burns and Shulgan, *Autonomy,* 3–11.

8 **$80 billion surged into self-driving vehicle technologies**: Cameron F. Kerry and Jeff Karsten, "Report: Gauging Investment in Self-Driving Cars," Brookings Institution, October 16, 2017, https://www.brookings.edu/research/gauging-investment-in-self-driving-cars/.

8 **first truly self-driving taxi service, in Chandler, Arizona**: John Krafcik, "Waymo One: The Next Step on Our Self-Driving Journey," *Waymo* (blog), Medium, December 5, 2018, https://medium.com/waymo/waymo-one-the-next-step-on-our-self-driving-journey-6d0c075b0e9b.

8 **set aside more than $10 billion**: "Waymo CEO on Future of Autonomous Vehicles," *Bloomberg*, video, September 13, 2017, https://www.bloomberg.com/news/videos/2017-09-14/waymo-ceo-on-future-of-autonomous-vehicles-video.

8 **"There is hardly a task that horse-drawn vehicles"**: "Practical Automobiles: This Type to Form Large Portion of Coming Exhibition," *New York Times*, January 12, 1903, https://timesmachine.nytimes.com/timesmachine/1903/01/12/101965824.pdf.

9 **"I was ready to make a leap into the future"**: David Leonhardt, "Driverless Cars Made Me Nervous. Then I Tried One," Opinion, *New York Times*, October 22, 2017, https://www.nytimes.com/2017/10/22/opinion/driverless-cars-test-drive.html.

9 **60 million people were killed**: *Death: A Self-Portrait*, 2012, Richard Harris Collection, London, UK: Wellcome Collection, exhibition.

9 **time wasted in traffic**: "INRIX Global Traffic Scorecard," INRIX, accessed February 15, 2018, http://inrix.com/scorecard.

10 **25 million people have disabilities that limit travel**: Stephen Brumbaugh, *Travel Patterns of Americans with Disabilities* (Washington, DC: Bureau of Transportation Statistics, 2018), https://www.bts.gov/sites/bts.dot.gov/files/docs/explore-topics-and-geography/topics/passenger-travel/222466/travel-patterns-american-adults-disabilities-9-6-2018_1.pdf.

10 **By 2030 . . . *tens of millions***: BlackRock Investment Group, *Future of the Vehicle: Winners and Losers: From Cars and Cameras to Chips* (BlackRock Investment Institute, 2017), 8.

10 **two billion human-driven cars and trucks**: Daniel Sperling and Deborah Gordon, *Two Billion Cars: Driving toward Sustainability* (New York: Oxford University Press, 2009).

10 **"The future is already here"**: Marianne Trench, *Cyberpunk* (New York: Inter-con Production, 1990), YouTube video.

11 **a mix of both worlds**: Bern Grush and John Niles, *The End of Driving: Transportation Systems and Public Policy Planning for Autonomous Vehicles* (Cambridge, MA: Elsevier, 2018).

11 **Half will end up in China**: "Autonomous Vehicle Sales to Surpass 33 Million Annually in 2040, Enabling New Autonomous Mobility in More Than 26 Percent of New Car Sales, IHS Markit Says," IHS Markit, January 2, 2018, https://technology.ihs.com/599099/autonomous-vehicle-sales-to-surpass -33-million-annually-in-2040-enabling-new-autonomous-mobility-in-more -than-26-percent-of-new-car-sales-ihs-markit-says.

11 **$2 trillion global auto-manufacturing industry**: Roger Lanctot, *Accelerating the Future: The Economic Impact of the Emerging Passenger Economy* (Strategy Analytics, June 2017), https://newsroom.intel.com/newsroom/wp -content/uploads/sites/11/2017/05/passenger-economy.pdf; **roughly the size of the entire EU economy**: "The Economy," European Union (website), accessed April 11, 2019, https://europa.eu/european-union/about-eu/figures/ economy_en.

11 **capture a $1.7 trillion annual share by 2030**: Author's calculation based on Peter Campbell, "Waymo Forecast to Capture 60% of Driverless Market," *Financial Times*, May 10, 2018, https://www.ft.com/content/3355f5b0-539d -11e8-b24e-cad6aa67e23e.

11 **of many shapes and sizes will have replaced them**: Scott Corwin et al., "The Future of Mobility: What's Next?" *Insights*, Deloitte, September 14, 2016, https://www2.deloitte.com/insights/us/en/focus/future-of-mobility/ roadmap-for-future-of-urban-mobility.html.

12 **cost of traffic congestion grew tenfold**: Bruce Schaller, "What Urban Sprawl Is Really Doing to Your Commute," *CityLab*, September 4, 2019, https:// www.citylab.com/perspective/2019/09/worst-cities-traffic-congestion -commuting-time-transit-data/597262/.

12 **car was cheap, rugged, and simple to repair**: Joshua B. Freeman, *Behemoth: A History of the Factory and the Making of the Modern World* (New York: W. W. Norton, 2017), 117–19.

13 **adoption of municipal zoning laws across America in the 1920s**: William A. Fischel, "An Economic History of Zoning and a Cure for its Exclusionary Effects," *Urban Studies* 41 (2004): 317–40.

15 **"a range of strong policies to achieve"**: Lew Fulton et al., *Three Revolutions in Urban Transportation* (Davis, CA: Institute of Transportation Studies, 2017).

2. Deconstructing Driving

21 **"In future commutes, you won't have to focus"**: Trefer Moss and Lisa Lin, "Don't Call It a Car: China's Internet Giants Want to Sell You 'Mobile Living Spaces,'" *Wall Street Journal*, March 18, 2018, https://www.wsj.com/articles/now-chinas-internet-giants-are-shaking-up-the-car-industry-1521374401.

22 **about 70 percent of 17-year-olds**: Brandon Schoettle and Michael Sivak, *Recent Changes in the Age Composition of U.S. Drivers: Implications for the Extent, Safety, and Environmental Consequences of Personal Transportation* (Ann Arbor: University of Michigan Transportation Research Institute, 2011), http://dx.doi.org/10.1080/15389588.2011.605817.

22 **A similar slide shows up**: Brandon Schoettle and Michael Sivak, *Recent Changes in Age Composition of Drivers in 15 Countries* (Ann Arbor: University of Michigan Transportation Research Institute, 2011), http://hdl.handle.net/2027.42/86680.

22 **90 percent of American teens . . . had access to a smartphone**: Monica Anderson and Jinjing Jiang, "Teens, Social Media and Technology 2018," Pew Research Center, May 31, 2018, https://www.pewinternet.org/2018/05/31/teens-social-media-technology-2018/; **had access to a computer or game console**: D'vera Cohn, "Census: Computer Ownership, Internet Connection Varies Widely across U.S.," Pew Research Center, September 19, 2014, http://www.pewresearch.org/fact-tank/2014/09/19/census-computer-ownership-internet-connection-varies-widely-across-u-s/.

22 **"For some, Mom and Dad are such good chauffeurs"**: Jean M. Twenge, "Have Smartphones Destroyed a Generation?" *The Atlantic*, September 2017, https://www.theatlantic.com/magazine/archive/2017/09/has-the-smartphone-destroyed-a-generation/534198/.

23 **widespread shift to *graduated licensing***: Tim Henderson, "Why Many Teens Don't Want to Get a Driver's License," *PBS NewsHour*, March 6, 2017, https://www.pbs.org/newshour/nation/many-teens-dont-want-get-drivers-license.

23 **the rate of decline in licensing was already slowing down**: According to a 2016 University of Michigan study, in the US the share of 16-year-olds with a driver's license fell by 47 percent between 1983 and 2014. But the bulk of that drop occurred before 2011, after which there was just a 10.9 percent decline in the next three years. See Michael Sivak and Brandon Schoettle, *Recent Decreases in the Proportion of Persons with a Driver's License across Age Groups* (Ann Arbor: University of Michigan Transportation Research Institute, 2013), http://umich.edu/~umtriswt/PDF/UMTRI-2016-4.pdf.

23 **the typical American car buyer today is more than 50 years old**: "Car Buyers Getting Older, Richer, NADA Economist Says," *Automotive News*, August 4, 2015, https://www.autonews.com/article/20150804/RETAIL03/150809938/car-buyers-getting-older-richer-nada-economist-says.

24 **"even then, sometimes the driver was just weary"**: Norm Nyhuis, "Drowsy Driving," Evergreen Safety Council, November 2008, http://www.esc.org/wp-content/uploads/2008-11-November-Newsletter-2008-rev2.pdf.

24 **35-mile-per-hour national speed limit**: Jim Donnelly, "Ralph R. Teetor," *Hemmings Classic Car*, July 2009, https://www.hemmings.com/magazine/hcc/2009/07/Ralph-R--Teetor/1846418.html.

24 **who coined their own ungainly monikers**: "Ralph Teetor and the History of Cruise Control," blog post, American Safety Council, accessed May 7, 2018, http://blog.americansafetycouncil.com/history-of-cruise-control-2/.

25 **drowsy driving occurred *more* frequently**: VINCI Autoroutes Foundation, "Cruise Control and Speed Limiters Impact Driver Vigilance," press release, July 30, 2013, https://fondation.vinci-autoroutes.com/fr/system/files/pdf/2013/07/pr_vinci_autoroutes_foundation_cruise_control_and_speed_limiters_impact_.pdf.

25 **analyzed telemetry from more than 270 million car trips**: Keith Barry, "Drivers More Likely to Use Their Phones When Cruise Control Is On, Study Finds," *Consumer Reports*, May 30, 2018, https://www.consumerreports.org/car-safety/drivers-more-likely-to-use-phones-when-cruise-control-is-on-mit-study/.

27 **Autopilot "sees"**: "Software Version 9.0," Tesla, accessed October 1, 2019, https://www.tesla.com/support/software-v9.

27 **Autopilot routinely lets drivers**: Patrick Olsen, "CR Finds That These Features Making Driving Easier but Introduce New Safety Risks," *Consumer Reports*, October 4, 2018, https://www.consumerreports.org/autonomous-driving/cadillac-tops-tesla-in-automated-systems-ranking/.

27 **a startling number of bugs**: Liane Yvkoff, "Is Tesla's Autopilot Seeing Ghosts?" *The Drive*, August 1, 2016, https://www.thedrive.com/news/4670/is-teslas-autopilot-seeing-ghosts.

28 **mistakenly classified by Autopilot**: Andrew J. Hawkins, "Tesla's Autopilot Was Engaged When Model 3 Crashed into Truck, Report States," *The Verge*, May 16, 2019, https://www.theverge.com/2019/5/16/18627766/tesla-autopilot-fatal-crash-delray-florida-ntsb-model-3.

28 **Brown spent a mere 25 seconds of the final 37 minutes of the trip**: National Transportation Safety Board, *Collision between a Car Operating with Automated*

Vehicle Control Systems and a Tractor-Semitrailer Truck near Williston, Florida, Accident Report NTSB/HAR-17/02, PB2017-102600, October 12, 2017, 15.

28 **A single light touch**: National Transportation Safety Board, *Collision between a Car*, 11.

28 **an audible alert after 60 seconds**: David Shepardson, "Tesla, Others Seek Ways to Ensure Drivers Keep Their Hands on the Wheel," *Reuters*, last modified June 23, 2017, https://www.reuters.com/article/us-usa-autos-selfdriving -safety/tesla-others-seek-ways-to-ensure-drivers-keep-their-hands-on-the -wheel-idUSKBN19E1ZA.

29 **riding in the passenger seat**: Telegraph Reporters, "Tesla Owner Who Turned On Car's Autopilot Then Sat in Passenger Seat While Travelling on the M1 Banned from Driving," *The Telegraph*, April 28, 2018, https://www.telegraph .co.uk/news/2018/04/28/tesla-owner-turned-cars-autopilot-sat-passenger -seat-travelling/.

29 **The intoxicated driver was found passed out**: Doug Smith, "CHP Uses Autopilot to Stop a Tesla Model S with a Sleeping Driver at the Wheel," *Los Angeles Times*, December 3, 2018, https://www.latimes.com/local/lanow/la-me-ln -tesla-driver-asleep-20181202-story.html.

29 **"Because of the impressive ability of Tesla's Autopilot"**: Patrick Olsen, "CR Finds That These Features Making Driving Easier but Introduce New Safety Risks," *Consumer Reports*, October 4, 2018, https://www.consumerreports .org/autonomous-driving/cadillac-tops-tesla-in-automated-systems -ranking/.

29 **one-third of the three-hour trip looking away**: Hod Lipson and Melba Kurman, *Driverless: Intelligent Cars and the Road Ahead* (Cambridge, MA: MIT Press, 2018), 60–61.

29 **"pointed at the driver's face"**: Alex Roy, "The Half-Life of Danger: The Truth behind the Tesla Model X Crash," *The Drive*, April 16, 2018, http://www .thedrive.com/opinion/20082/the-half-life-of-danger-the-truth-behind-the -tesla-model-x-crash.

29 **after 15 seconds it disengages**: Jonathan M. Gitlin, "GM Rolling Out Its Amazing Super Cruise Tech to More Cars and Brands," *Ars Technica*, June 6, 2018, https://arstechnica.com/cars/2018/06/butt-kicking-super-cruise-com ing-to-all-my2020-cadillacs-more-gms-later/.

29 **an entire science of "crew resource management"**: Federal Aviation Administration, "The History of CRM," *FAA TV*, 24:16, April 5, 2012, https:// www.faa.gov/tv/?mediaId=447.

30 **how we spend the time . . . has shifted**: Chen Song and Chao Wei, "Travel

Time Use over Five Decades," *Transportation Research Part A: Policy and Practice* 116 (October 2018): 73–96.

30 **Audi's concept car of the future**: Chris Paukert, "Audi's Long Distance Lounge Hypes a Smarter Autonomous Future," *Roadshow, CNET,* June 14, 2017, https://www.cnet.com/roadshow/news/audi-long-distance-lounge-au tonomous-concept-exclusive-hands-on-video/.

30 **self-driving vehicles could free up 250 million hours**: Roger Lanctot, *Accelerating the Future: The Economic Impact of the Emerging Passenger Economy* (Strategy Analytics, June 2017), 6, https://newsroom.intel.com/newsroom/wp -content/uploads/sites/11/2017/05/passenger-economy.pdf.

30 **$150 billion in the US alone**: Securing America's Future Energy, *America's Workforce and the Self-Driving Future: Realizing Productivity Gains and Spurring Economic Growth*, June 2018, 22, https://avworkforce.secureenergy.org/ wp-content/uploads/2018/06/SAFE_AV_Policy_Brief.pdf.

31 **asked how they expect to spend their saved time**: Eva Fraedrich et al., *User Perspectives on Autonomous Driving: A Use-Case-Driven Study in Germany* (Berlin, Germany: DLR Institute of Transport Research, 2016), 13; Chris Tennant et al., Executive Summary, *Autonomous Vehicles—Negotiating a Place on the Road* (London, UK: London School of Economics, 2016), 1–10.

31 **"Why do all of these interior designs"**: Alanis King, "Autonomous Cars Aren't Even Here Yet and I'm Already Bored with Them," *Jalopnik*, September 11, 2017, https://jalopnik.com/autonomous-cars-arent-even-here-yet-and-im -already-bore-1803756153.

31 **Audi has recruited Disney**: Reese Counts, "We Try Audi and Disney's New In-Car Entertainment System on the Track," *Autoblog,* January 9, 2019, https:// www.autoblog.com/2019/01/09/audi-disney-holoride-car-vr-entertainment/.

31 **Kia built a concept car**: Laura Bliss, "The 'Driverless Experience' Looks Awfully Distracting," *CityLab*, January 11, 2019, https://www.citylab.com/ transportation/2019/01/self-driving-car-technology-consumer-electronics -show/580027/.

31 **bigger . . . than the entire auto industry today**: Lanctot, *Accelerating the Future*, 5.

31 **serve up precision-targeted media:** Joann Muller, "One Big Thing: What Your Car Will Know about You," *Axios*, May 10, 2019, https://www.axios .com/newsletters/axios-autonomous-vehicles-7b382e7a-e9f1-466b-9c7b -33e4aadc03f4.html; **sensors. . . uniquely identify your heartbeat**: "Goode Intelligence Forecasts That Biometrics Market for the Connected Car Will Be Just under $1bn by 2023," *Goode Intelligence,* November 13, 2017, https://www w

.goodeintelligence.com/wp-content/uploads/2017/11/Goode-Intelligence -Biometrics-for-the-Connected-Car_Nov17_-news_release-13112017.pdf.

32 **GM already tracks:** Jamie LaReau, "GM Tracked Radio Listening Habits for 3 Months: Here's Why," *Detroit Free Press*, October 1, 2018, https://www .freep.com/story/money/cars/general-motors/2018/10/01/gm-radio-listening -habits-advertising/1424294002/.

32 **"We know how long they've lived":** Phoebe Wall Howard, "Data Could Be What Ford Sells Next as It Looks for New Revenue," *Detroit Free Press*, November 13, 2018, https://www.freep.com/story/money/cars/2018/11/13/ford -motor-credit-data-new-revenue/1967077002/.

33 **odds of a crash instantly double:** National Highway Traffic Safety Administration, *Overview of the National Highway Traffic Safety Administration's Driver Distraction Program*, DOT HS 811 299, April 2010, https://www.nhtsa.gov/ sites/nhtsa.dot.gov/files/811299.pdf.

33 **road crashes in the US declined:** Wikipedia, s.v. "Motor Vehicle Fatality Rate in U.S. by Year," accessed April 10, 2019, https://en.wikipedia.org/wiki/Motor_ vehicle_fatality_rate_in_U.S._by_year.

33 **deaths from distracted driving:** Fernando A. Wilson and Jim P. Stimpson, "Trends in Fatalities from Distracted Driving in the United States, 1999 to 2008," *American Journal of Public Health* 100, no. 11 (2010): 2213–19, http://doi .org/10.2105/AJPH.2009.187179.

33 **400,000 injuries were blamed on distracted driving:** National Highway Traffic Safety Administration, *2016 Fatal Motor Vehicle Crashes: Overview*, Traffic Safety Facts Research Note, DOT HS 812 456, October 2017, https:// crashstats.nhtsa.dot.gov/Api/Public/ViewPublication/812456.

33 **four times more likely to be in a crash:** World Health Organization, *Mobile Phone Use: A Growing Problem of Driver Distraction* (Geneva, Switzerland: WHO Publications, 2011), 3.

33 **strict laws restricting the use of mobile devices:** National Highway Traffic Safety Administration, "Distracted Driving Global Fact Sheet," accessed January 31, 2019, https://usdotblog.typepad.com/files/6983_distracteddrivingfs_5 -17_v2.pdf.

34 **360-degree field of view:** Waymo, *Waymo 360° Experience: A Fully Self-Driving Journey*, February 2018, YouTube video, https://www.youtube.com/ watch?v=B8R148hFxPw.

35 **reduced the cost of lidar:** Kirsten Korosec, "Five Things to Know about the Future of Google's Self-Driving Car Company: Waymo," *Fortune*, January 8, 2017, http://fortune.com/2017/01/08/waymo-detroit-future/.

35 **higher resolution than radar and longer range**: Sebastian Thrun et al., "Stanley: The Robot That Won the DARPA Grand Challenge," *Journal of Field Robotics* 23, no. 9 (2006): 661–92, http://isl.ecst.csuchico.edu/DOCS/darpa2005/DARPA%202005%20Stanley.pdf.

35 **more than 3,000 smartphone-toting citizens**: Author's calculation based on Brian Krzanich, "Data Is the New Oil in the Future of Automated Driving," Intel, November 15, 2016, https://newsroom.intel.com/editorials/krzanich-the-future-of-automated-driving/#gs.dcqfk7.

36 **neural networks were put to work in banks and postal systems**: Leon Bottou, "Graph Transformer Networks," Leon Bottou (website), September 28, 2018, https://leon.bottou.org/talks/gtn.

37 **The occupancy grid provides a geometric structure**: Lipson and Kurman, *Driverless*, 93–94.

38 **these predictions are represented by a *cone of uncertainty***: Lipson and Kurman, *Driverless*, 95–98.

38 **more people will have died at the hands of drivers**: World Health Organization, *Global Status Report on Road Safety: Time for Action* (Geneva, Switzerland: World Health Organization, 2009), ix.

38 **kill more children and young adults (age 5 to 29)**: World Health Organization, *Global Status Report on Road Safety 2018* (Geneva, Switzerland: World Health Organization, 2018).

40 **autonomous vehicles can operate**: For instance, see Lipson and Kurman, *Driverless*, 127–36, 143–48, which scopes out a strictly limited role for government in the driverless revolution.

40 **"Why would we invest in putting wires in the road?"**: Dan Albert, *Are We There Yet? The American Automobile Past, Present, and Driverless* (New York: W. W. Norton, 2019), 253.

40 **efforts by libertarian think tanks**: Hubert Horan, "Uber's Path of Destruction," *American Affairs* 3, no. 2 (Summer 2019), https://americanaffairsjournal.org/2019/05/ubers-path-of-destruction/.

40 **autonomous-driving software are still trying to understand**: Wendy Ju, interview with author, June 14, 2018.

40 **Data from AVs in the field**: Adam Grzywaczewski, "Training AI for Self-Driving Vehicles: The Challenge of Scale," *Nvidia Developer Blog*, October 9, 2017, https://devblogs.nvidia.com/training-self-driving-vehicles-challenge-scale/.

41 **300,000 online gig workers of Seattle-based Mighty AI**: "Technology: From Here to Autonomy," in "Special Report: Autonomous Vehicles," *The Economist*, March 3, 2018, 3–5.

41 **fleet of Chevy Bolts disengaged on average once every 1,254 miles**: "Technology: From Here to Autonomy," 3–5.

42 **Waymo managed just a 10 percent improvement in the rate**: Andrew J. Hawkins, "Waymo and GM Still Lead the Pack in California's New Self-Driving Report Cards," *The Verge*, January 31, 2018, https://www.theverge.com/2018/1/31/16956902/california-dmv-self-driving-car-disengagement-2017.

42 **more than doubling the average distance between disengagements**: Alan Ohnsman, "Waymo Tops Self-Driving Car 'Disengagement' Stats as GM Cruise Gains and Tesla Is AWOL," *Forbes*, February 13, 2019, https://www.forbes.com/sites/alanohnsman/2019/02/13/waymo-tops-self-driving-car-disengagement-stats-as-gm-cruise-gains-and-tesla-is-awol/#7b83615131ec.

42 **"freedom from external control or influence"**: *English Oxford Living Dictionaries*, s.v. "Autonomy," accessed February 2, 2019, https://en.oxforddictionaries.com/definition/us/autonomy.

42 **"has the capability to drive a vehicle without"**: "Key Autonomous Vehicle Definitions," State of California Department of Motor Vehicles, accessed February 1, 2019, https://www.dmv.ca.gov/portal/dmv/detail/vr/autonomous/definitions.

42 **400,000 "fifth-generation" (5G) wireless sites**: Dan Jones, "5G: The Density Question," *Light Reading*, February 15, 2018, https://www.lightreading.com/mobile/5g/5g-the-density-question-/a/d-id/740634?.

42 **$150 billion or so that's needed to pay for it**: "Deep Deployment of Fiber Optics Is a National Imperative," Deloitte, accessed February 1, 2019, https://www2.deloitte.com/us/en/pages/consulting/articles/communications-infrastructure-upgrade-deep-fiber-imperative.html.

43 **GM embraced a vision of joint "national purpose"**: Albert, *Are We There Yet*, 246.

44 **socially acceptable standard of cleanliness rose**: Ruth Cowan, *More Work for Mother: The Ironies of Household Technology from the Open Hearth to the Microwave* (New York: Basic Books, 1985).

45 **"passengers will still pay extra for a better driving"**: "Last Lap of Luxury: German Cars Have the Most to Lose from a Changing Auto Industry," *The Economist*, March 2018, 57.

45 **Pilots of today's aircraft now require**: Hillary Abraham et al., "What's in a Name: Vehicle Technology Branding and Consumer Expectations for Automation," paper presented at AutomotiveUI '17, Oldenburg, Germany, September 2017, https://doi.org/10.1145/3122986.3123018.

45 **"Technology does not eliminate error"**: Vikas Bajaj, "The Bright, Shiny Distraction of Self-Driving Cars," Opinion, *New York Times*, March 31, 2018, https://www.nytimes.com/2018/03/31/opinion/distraction-self-driving-cars.html.

45 **The further the spread of automation**: David H. Keller, "The Living Machine," *Wonder Stories*, May 1935, 1471.

45 **the rate of skepticism had increased to 73 percent**: "AAA: American Trusts in Autonomous Vehicles Slips," Newsroom, AAA, May 22, 2018, https://newsroom.aaa.com/2018/05/aaa-american-trust-autonomous -vehicles-slips/.

46 **"rider-only" taxi service**: Joe White, "Waymo Tests 'Rider Only' Service and Looks Beyond Robo-taxis," *Reuters*, October 28, 2019, https://www.reuters .com/article/us-autos-selfdriving-waymo-idUSKBN1X71U7.

46 **approve AV-planned detours**: Ellice Perez, "Getting Ready for More Early Riders in Phoenix," *Waymo* (blog), Medium, August 21, 2018, https://medium .com/waymo/getting-ready-for-more-early-riders-in-phoenix-1699285cbb84.

46 **the way harbor pilots take control of big ships**: Darrell Etherington, "Starsky Robotics' Autonomous Transport Trucks Also Give Drivers Remote Control," *TechCrunch*, February 28, 2017, https://techcrunch.com/2017/02/28/ starsky-robotics-autonomous-transport-trucks-also-give-drivers-remote -control/.

46 **MIT's Sangbae Kim builds robots that move over terrain**: Cara Giaimo, "Forces of Nature," *MIT Technology Review*, December 19, 2018, https://www .technologyreview.com/s/612527/forces-of-nature/.

3. The Origin of (Vehicular) Species

49 **"We might need self-driving buildings as well as self-driving cars"**: National League of Cities, "Urban Transformation: Reprogramming Buses, Bikes, and Barriers," *Autonomous Vehicles: Future Scenarios*, accessed April 12, 2019, http://avfutures.nlc.org/urban-transformation.

49 **so comfortable with the absence of change**: William Samuelson and Richard Zeckhauser, "Status Quo Bias in Decision Making," *Journal of Risk and Uncertainty* 1, no. 1 (1988): 7–59.

50 **"No traffic jams . . . no collisions . . . no driver fatigue"**: America's Independent Electric Light and Power Companies, "Advertising," *Life*, January 1956, 8.

54 **"to buy coffee in the morning, to drop off"**: Allan E. Pisarski, "Commuting in America," *Issues in Science and Technology*, Winter 2007, https://issues.org/ realnumbers-22/.

54 **factories today are the largest humankind has ever built**: Joshua B. Freeman, *Behemoth: A History of the Factory and the Making of the Modern World* (New York: W. W. Norton, 2017), xiii.

55 **AV's electric motors and batteries could be downsized**: William J. Mitchell, Christopher E. Borroni-Bird, and Lawrence D. Burns, *Reinventing the Automobile: Personal Urban Mobility for the 21st Century* (Cambridge, MA: MIT Press, 2010), 24.

56 **"up to four bags of groceries"**: "Marble's Delivery Robots Hit the Ground to Map Out Arlington, Texas," AUVSI, August 21, 2018, https://www .auvsi.org/industry-news/marbles-delivery-robots-hit-ground-map-out -arlington-texas.

57 **a global web of design and engineering, manufacturing**: "Starship Technologies," *Crunchbase*, accessed February 11, 2019, https://www.crunchbase .com/organization/starship-technologies#section-investors.

57 **the city's Board of Supervisors to enact a temporary ban**: Adam Brinklow, "San Francisco Ready to Permit Robots on City Sidewalks," *Curbed SF*, March 14, 2018, https://sf.curbed.com/2018/3/14/17120628/san-francisco-robot-ban -fees-yee-tech.

57 **"intricate sidewalk ballet"**: Jane Jacobs, *The Death and Life of Great American Cities* (New York: Random House, 1961), 92.

57 **a conveyor with enough sense to learn our unwritten rules**: Jennifer Chu, "New Robot Rolls with Rules of Pedestrian Conduct," *MIT News*, Massachusetts Institute of Technology, August 29, 2017, http://news.mit.edu/2017/new -robot-rolls-rules-pedestrian-conduct-0830.

57 **"students held a candlelight vigil for it"**: Carolyn Said, "Kiwibots Win Fans at UC Berkeley as They Deliver Fast Food at Slow Speeds," *San Francisco Chronicle*, May 26, 2019, https://www.sfchronicle.com/business/article/Kiwibots -win-fans-at-UC-Berkeley-as-they-deliver-13895867.php.

58 **capping a sudden surge of venture-capital investment**: Leslie Hook and Tim Bradshaw, "Silicon Valley Start-Ups Race to Win in Driverless Delivery Market," *Financial Times*, January 30, 2018, https://www.ft.com/content/ decb36d8-056b-11e8-9650-9c0ad2d7c5b5.

58 **Ford plans to launch one based on a pickup truck chassis**: Joann Muller, "One Big Thing: Ford CEO Says AV Progress Isn't All about Technology," *Axios*, October 12, 2018, https://www.axios.com/newsletters/axios-autonomous -vehicles-d682a456-7a3d-4e62-bcf0-5b322f82df48.html.

58 **5.14-meter-long vehicle can carry 12 passengers**: "Vision Urbanetic: On Demand, Efficient and Sustainable: Vision Urbanetic Answers the Question of

Future Urban Mobility," Daimler, September 10, 2018, https://media.daimler
.com/marsMediaSite/en/instance/ko.xhtml?oid=41169541.

63 **deal with the diminutive nation-state of Andorra**: "An Alternative Autonomous Revolution," MIT Media Lab, accessed February 11, 2019, https://www
.media.mit.edu/projects/pev/overview/.

63 **millions of shared bikes serve a thousand cities**: Renate van der Zee, "Story
of Cities #30: How This Amsterdam Inventor Gave Bike-Sharing to the World,"
The Guardian, April 26, 2016, https://www.theguardian.com/cities/2016/
apr/26/story-cities-amsterdam-bike-share-scheme.

63 **Provo launched an experiment**: Van der Zee, "Story of Cities."

64 **the number of dockless bikes in Shanghai**: Felix Salmon, "Bring on the
Bikocalypse," *Wired*, February 1, 2018, https://www.wired.com/story/chinese
-dockless-bikes-revolution/.

66 **this miniature AV is expected to offset its high initial cost**: Yingzhi Zhang
and Brenda Goh, "China's Ninebot Unveils Scooters That Drive Themselves to
Charging Stations," *Reuters*, August 16, 2019, https://ca.reuters.com/article/
technologyNews/idCAKCN1V60LJ-OCATC.

67 **the elderly and nonambulatory will be huge markets**: "Autonomous Wheelchairs," Hitachi, accessed May 19, 2018, http://www.hitachi.com/rd/portal/
highlight/vision_design/future/autonomous_mobility/chair/index.html.

67 **Americans in wheelchairs grew by half between 2010 and 2014**: Henry
Claypool, "An Aging Population Could Increase Demand for Autonomous
Vehicles," *Axios*, July 12, 2019, https://www.axios.com/an-aging-population
-could-increase-demand-for-autonomous-vehicles-b6a9b58e-b073-4151
-98bf-5365a5e5ada0.html.

67 **Segway's S-Pod:** Sean O'Kane, "Segway's S-Pod looks weird, but it's a lot of
fun to drive," *The Verge*, January 8, 2020, https://www.theverge.com/2020/1/8
/21056268/segway-s-pod-first-drive-hands-on-ces-2020.

67 **a self-balancing electric motorcycle**: Scott C. Wiley, Self-Balancing Robotic
Motorcycle, US Patent application 828,387, publication May 30, 2019, http://
www.freepatentsonline.com/20190161132.pdf.

67 **recovers the up-front cost of its scooters**: Haje Jan Kamps, "How to Understand the Financial Levers in Your Business, or How Can an Electric Scooter
Ride-Sharing Company Like Bird Possibly Make Money?" blog post, *Bolt*, April
16, 2018, https://blog.bolt.io/financial-models-faaece0871bc.

67 **a more durable scooter model that carries built-in maintenance sensors:** Andrew J. Hawkins, "Bird's New Electric Scooter Has Better Battery and
Anti-vandalism Sensors," *The Verge*, August 1, 2019, https://www.theverge

.com/2019/8/1/20749511/bird-two-electric-scooter-battery-autonomous
-sensors.

67 **nine versions of its flagship bike**: Michal Naka (@michalnaka), "Lime has
been through 9 vehicle versions in 17 months of existence," Twitter, May 9,
2018, 4:20 a.m., https://twitter.com/michalnaka/status/994235665377181698.

67 **new design idea into on-street hardware in 15 days**: "VeoRide Leads
Micromobility Industry with Quick, Continuous Improvements," *PR News-
wire*, March 27, 2019, https://www.prnewswire.com/news-releases/veoride
-leads-micromobility-industry-with-quick-continuous-improvements
-300819663.html.

68 **a liquid-cooled supercomputer that packed more power per ounce**: David
H. Freedman, "Self-Driving Trucks: Ten Breakthrough Technologies 2017,"
MIT Technology Review, February 22, 2017, https://www.technologyreview
.com/s/603493/10-breakthrough-technologies-2017-self-driving-trucks/.

68 **"It's like a train on software rails"**: Kelsey Atherton, "Uber Freight Goes
After the Trucking Business," *Popular Science*, December 27, 2016, https://
www.popsci.com/uber-freight-takes-aim-at-trucking.

69 **Scania successfully tested platoons**: Scania Group, "Semi-autonomous
Truck Platooning—How Does It Work?" YouTube video, January 29, 2018,
https://www.youtube.com/watch?v=lpuwG4A56r0.

69 **"road trains" are still commonplace**: The longest road train in history was
more than 4,800 feet long and consisted of 113 trailers. Assembled in 2006 for
an event at the Hogs Breath Café, in Clifton, Queensland, Australia, the over-
sized hauler heaved just 490 feet in a little under one minute before securing
its spot in the Guinness World Records. "Longest road train," *Guinness World
Records*, accessed February 11, 2019, http://www.guinnessworldrecords.com/
world-records/longest-road-train.

69 **efficiency gains offered by platooning weren't enough**: "Daimler Trucks
Invests Half a Billion Euros in Highly Automated Trucks," Daimler, January 7,
2019, https://media.daimler.com/marsMediaSite/ko/en/42188247.

69 **more than 33 million people in 195 cities worldwide**: Ryan Winstead, "BRT
Hits 400 Corridors and Systems Worldwide," *The City Fix*, August 31, 2015,
http://thecityfix.com/blog/brt-hits-400-corridors-systems-worldwide-ryan
-winstead/.

69 **half or less than the throughput of the world's busiest subways**: Jeffrey
Ng, "Hong Kong's Subway System Wants to Rule the World," *Wall Street Journal*, September 18, 2013, https://www.wsj.com/articles/hong-kongs-mtr-rides
-toward-expansion-1379436218.

72 **$200 million to an AV system it calls CityPilot**: Chris O'Brien, "Robotic Buses Leapfrog Self-Driving Trucks in Autonomy Revolution," *Trucks.com*, February 27, 2017, https://www.trucks.com/2017/02/27/buses-european-self -driving-vehicle-revolution/.

72 **"It's much closer than many people realize"**: Daniel Salazar, "Lots of Driverless Buses with Dedicated Lanes: How Capital Metro Sees Austin's Public Transit Future," *Austin Business Journal*, October 3, 2018, https://www .bizjournals.com/austin/news/2018/10/03/lots-of-driverless-buses-with -dedicated-lanes-how.html.

73 **"But in a future where people no longer have"**: Anne Quito, "IKEA's Think Tank Envisions Self-Driving Cars as Rooms on Wheels," *Quartz*, September 18, 2018, https://qz.com/quartzy/1392507/ikea-funded-future-living-lab -space10-prototypes-autonomous-vehicles-for-public-services/.

74 **"the international icon of radical architecture of the Sixties"**: Sutherland Lyall, "Obituary: Ron Herron," *The Independent*, October 5, 1994, https:// www.independent.co.uk/news/people/obituary-professor-ron-herron -1440981.html.

74 **weighs some 76 tons and carries up to 90 people**: Michael J. Gaynor, "Why Do We Still Have Those Weird-Looking People Movers at Dulles?" *Washingtonian*, November 21, 2016, https://www.washingtonian.com/2016/11/21/ mobile-lounge-dulles-airport-people-movers/.

74 **the 12-mile journey required a custom-built 160-wheel carrier**: "Space Shuttle Endeavour's Trek across LA: Timelapse," YouTube video, October 19, 2012, https://www.youtube.com/watch?v=JdqZyACCYZc; **and hundreds of human escorts, at a cost of more than $10 million**: "Space Shuttle Endeavour's Move May Cost More Than $10 Million," *L.A. Now* (blog), *Los Angeles Times*, October 12, 2012, https://latimesblogs.latimes .com/lanow/2012/10/space-shuttle-endeavours-move-to-cost-more-than -10-million.html.

79 **adorable droid trundles along footpaths**: "Humanizing Public Safety," Hitachi, accessed February 11, 2019, http://www.hitachi.com/rd/portal/ highlight/vision_design/future/city_home/green/safe/index.html.

79 **"The robotization of the city"**: National League of Cities, "Urban Transformation."

79 **MIT's Cheetah 3 is superior to many AVs**: Rima Sabina Aouf, "MIT's Blind Cheetah 3 Robot Can Navigate without Sensors or Cameras," Dezeen, July 17, 2018, https://www.dezeen.com/2018/07/17/institute-of-technology-mit -blind-cheetah-3-robot massachusetts-technology/.

80 **started on self-driving tractors**: Marina Lewycka, *A Short History of Tractors in Ukrainian* (London, UK: Viking Press, 2005).

80 **"a striking fact about the automobile"**: Brian Ladd, *Autophobia: Love and Hate in the Automotive Age* (Chicago: University of Chicago Press, 2008), 9.

4. Reprogramming Mobility

83 **"There are still millions of people who want to drive but lack the elemental courage"**: David H. Keller, "The Living Machine," *Wonder Stories*, May 1935, 1467.

83 **"injured little except his pride and clothing"**: Keller, "The Living Machine," 1465.

84 **"Something like a brain"**: Keller, "The Living Machine," 1469.

84 **we never learn what makes it tick**: Keller, "The Living Machine," 1469.

84 **"The blind for the first time were safe"**: Keller, "The Living Machine," 1467.

84 **Google's 2014 launch of a self-driving car prototype**: Google Self-Driving Car Project, "A First Drive," YouTube video, May 27, 2014, https://www.youtube .com/watch?v=CqSDWoAhvLU.

85 **"Traffic was paralyzed"**: Keller, "The Living Machine," 1510.

87 **all it takes is a tiny disruption to congeal the smooth flow**: Amiram Barkat, "OECD and IMF: Israel Has West's Worst Traffic Jams," *Globes*, March 14, 2018, https://en.globes.co.il/en/article-oecd-imf-israel-has-wests-worst -traffic-jams-1001227824.

88 **published a sweeping vision for future transportation**: Ashley Hand, *Urban Mobility in a Digital Age: A Transportation Technology Strategy for Los Angeles* (Los Angeles: Office of the Mayor and Department of Transportation, 2016), 1.

88 **Passenger travel on electric streetcars**: Charles W. Cheape, *Moving the Masses: Urban Public Transit in New York, Boston, and Philadelphia, 1880–1912* (Cambridge, MA: Harvard University Press, 1980), 174.

89 **as many as three sets of tracks and overhead power lines ran**: Cheape, *Moving the Masses*, 159.

89 **commuters paid at least two fares to get to work**: Cheape, *Moving the Masses*, 159.

89 **PTC's ridership had tripled:** Cheape, *Moving the Masses*, 174.

89 **the company's monopoly was secure**: Cheape, *Moving the Masses*, 162–67.

94 **"As a driverless taxi was finally introduced"**: Keller, "The Living Machine," 1470.

94 **drivers take home about 80 percent of ride-for-hire fees**: Alex Rosenblat, *Uberland: How Algorithms Are Rewriting the Rules of Work* (Oakland: University of California Press, 2018).

94 **all but eliminate the share that goes to labor**: International Transport Board of the OECD, *Urban Mobility System Upgrade: How Shared Self-Driving Cars Could Change City Traffic*, ITF Corporate Partnership Board Report, 2015.

94 **Today's cabs spend half of their working hours empty**: Judd Cramer and Alan B. Krueger, "Disruptive Change in Taxi Business: The Case of Uber," *American Economic Review* 106, no. 5 (2016): 177–82.

94 **could grow to $285 billion annually by 2030**: David Welch and Elisabeth Behrmann, "Who's Winning the Self-Driving Car Race?" *Bloomberg*, May 7, 2018, https://www.bloomberg.com/news/features/2018-05-07/who-s-winning -the-self-driving-car-race.

95 **more than *one billion* taxibots**: Author's calculation using estimates from UBS, *Longer Term Investments: Smart Mobility* (Chief Investment Office Americas, October 19, 2017).

95 **SilverRide targets senior citizens**: Mitchell Hartman, "Wanted: Elder Transportation Solutions," *Marketplace*, January 30, 2019, https://www.marketplace .org/2019/01/30/business/wanted-elder-transportation-solutions.

95 **"transported everything from leopards"**: Ted Trauter, "Pet Chauffeur Tried to Adapt to Tough Economy," *You're the Boss* (blog), *New York Times*, August 26, 2011, https://boss.blogs.nytimes.com/2011/08/26/pet-chauffeur-tries-to -adapt-to-tough-economy.

96 **no-cost rides to prenatal-care appointments and grocery stores**: Laura Bliss, "In Columbus, Expectant Moms Will Get On-Demand Rides to the Doctor," *CityLab*, December 27, 2018, https://www.citylab .com/transportation/2018/12/smart-city-columbus-prenatal-ride -hailing/579082/.

97 **the company could soon be serving up to a million passengers**: Alexis Madrigal, "Finally, the Self-Driving Car," *The Atlantic*, December 5, 2018, https:// www.theatlantic.com/technology/archive/2018/12/test-ride-waymos-self -driving-car/577378/.

97 **Singapore could make do with half**: MIT Senseable City Lab, "Unparking," Massachusetts Institute of Technology, accessed 20 February 2019, http:// senseable.mit.edu/unparking/.

97 **could swap one private self-driving cab for every six**: International Transport Board of the OECD, *Urban Mobility System Upgrade.*

98 **eliminate upwards of 75 percent of its yellow cabs**: Javier Alonso-Mora et al., "On-Demand High-Capacity Ride-Sharing via Dynamic Trip-Vehicle Assignment," *Proceedings of the National Academy of Sciences of the United States of America* 114, no. 3 (2017): 462–67.

98 **high cost of remote human safety monitors**: Ashley Nunes and Kristen Hernandez, "The Cost of Self-Driving Cars Will Be the Biggest Barrier to Their Adoption," *Harvard Business Review*, January 31, 2019, https://hbr.org/2019/01/the-cost-of-self-driving-cars-will-be-the-biggest-barrier-to-their-adoption.

98 **"We are going to also offer third-party"**: "Full Video and Transcript: Uber CEO Dara Khosrowshahi at Code 2018," *Recode*, June 4, 2018, https://www.recode.net/2018/5/31/17397186/full-transcript-uber-dara-khosrowshahi-code-2018.

99 **flubbed their geometry too**: For more, see Jarett Walker, "Does Elon Musk Understand Urban Geometry?" blog post, Human Transit, July 21, 2016, https://humantransit.org/2016/07/elon-musk-doesnt-understand-geometry.html.

99 **a catastrophic surge of traffic**: International Transport Board of the OECD, *Urban Mobility System Upgrade.*

99 **move nearly 60 percent of commuters**: "Is Informal Normal? Towards More and Better Jobs in Developing Countries," OECD, March 31, 2009, http://www.oecd.org/dev/inclusivesocietiesanddevelopment/isinformalnormaltowardsmoreandbetterjobsindevelopingcountries.htm.

100 **run more frequently with fewer breakdowns, than city buses**: Eric L. Goldwyn, "An Informal Transit System Hiding in Plain Sight: Brooklyn's Dollar Vans and Transportation Planning and Policy in New York City" (PhD diss., Columbia University, 2017), 96.

100 **hauling upwards of 35,000 workers on 800 vehicles**: Eillie Anzilotti, "The Reach of the Bay Area's 'Tech Buses,'" *CityLab*, September 16, 2016, https://www.citylab.com/transportation/2016/09/the-reach-of-the-bay-areas-tech-buses/500435/.

101 **"able to provide point-to-point service"**: Christopher Alexander, *A Pattern Language: Towns, Buildings, Construction* (New York: Oxford University Press, 1977), 112.

101 **gone back to the public-transit lines**: Laura Bliss, "Bridj Is Dead, but Microtransit Isn't," *CityLab*, May 3, 2017, https://www.citylab.com/transportation/2017/05/bridj-is-dead-but-microtransit-isnt/525156/.

102 **causing traffic and creating safety risks for pedestrians**: Georgios Kalog-

erakos, "Driverless Mobilities: Understanding Mobilities of the Future" (master's thesis, Aalborg University, 2017), 79.

102 **More than 12,000 people rode the six shuttles**: "City Overview: Trikala," CityMobil2, accessed June 13, 2018, http://www.citymobil2.eu/en/city -activities/large-scale-demonstration/trikala/ (site discontinued).

102 **more accepting of the technology than were townspeople in France**: City-Mobil2, *Experience and Recommendations* (CityMobil2, 2016), 32.

103 **a stylish name—the Navia**: Melissa Riofrio, "The Little Shuttle That Can: Induct Navia Is First Self-Driving Vehicle," *PC World*, January 8, 2014, https:// www.pcworld.com/article/2085006/the-little-shuttle-that-can-induct-navia -is-first-self-driving-vehicle.html.

103 **the company filed for bankruptcy**: Derek Christie et al., "Pioneering Driverless Electric Vehicles in Europe: The City Automated Transport Systems (CATS)," *Transportation Research Procedia* 13 (2016): 30–39, https://hal.inria .fr/hal-01357309/document.

103 **a completely redesigned vehicle**: "Sion, CH Is Piloting AVs," Initiative on Cities and Autonomous Vehicles, Bloomberg Philanthropies, accessed February 20, 2019, https://avsincities.bloomberg.org/global-atlas/europe/ch/sion -ch; "Project 'SmartShuttle,'" PostBus, accessed February 2019, https://www .postauto.ch/en/project-smartshuttle.

103 **driverless shuttles crawled along**: "These 61 Cities Are Piloting AVs for Transit," Initiative on Cities and Autonomous Vehicles, Bloomberg Philanthropies, accessed February 2019, https://avsincities.bloomberg.org/global-atlas/ tags/transit.

104 **ferried more than 1.5 million passengers**: National League of Cities, "Sustainability: Weaving a Microtransit Mesh," *Autonomous Vehicles: Future Scenarios*, accessed April 12, 2019, http://avfutures.nlc.org/sustainability.

104 **sold more than 100 Armas and went public**: "Navya Updates Its 2018 Revenue Target," Navya, December 7, 2018, https://navya.tech/en/press/navya -updates-its-2018-revenue-target/.

105 **cut operating costs by as much as 40 percent**: National League of Cities, "Sustainability: Weaving a Microtransit Mesh."

106 **a "curb kiss" fee**: Michael Cabanatuan and Kurtis Alexander, "Google Bus Backlash: S.F. to Impose Fees on Tech Shuttles," *SFGate*, January 21, 2014, https://www.sfgate.com/bayarea/article/Google-bus-backlash-S-F-to -impose-fees-on-tech-5163759.php.

107 **a planned driverless-shuttle network**: City of Bellevue, Washington, and

City of Kirkland, Washington, "A Flexible, Electric, Autonomous Commute-pool System," Bellevue-Kirkland USDOT (grant proposal, 2018).

108 **leftover data has value in predicting human behavior**: Shoshana Zuboff, *Surveillance Capitalism* (New York: Public Affairs, 2019).

109 **launched its own MaaS effort in 2019**: Adele Peters, "In Berlin, There's Now One App to Access Every Mode of Transportation," *Fast Company*, February 18, 2019, https://www.fastcompany.com/90308234/in-berlin-theres-now-one-app-to-access-every-mode-of-transportation.

109 **draws on a highly successful deployment in Vilnius**: Douglas Busvine, "From U-Bahn to E-Scooters: Berlin Mobility App Has It All," *Reuters*, September 24, 2019, https://www.reuters.com/article/us-tech-berlin/from-u-bahn-to-e-scooters-berlin-mobility-app-has-it-all-idUSKBN1W90MG.

110 **the role of mobility-service integrator**: Peters, "In Berlin."

110 **off to a slow start selling subscriptions**: Julia Walmsley, "Watch Out, Uber. Berlin Is the New Amazon for Transportation (with Lower Fares)," *Forbes*, June 26, 2019, https://www.forbes.com/sites/juliewalmsley/2019/06/26/watch-out-uber-berlin-is-the-new-amazon-for-transportation-with-lower-fares/#58cee4f9269b.

110 **legislated MaaS into existence**: David Zipper, "Helsinki's MaaS App, Whim: Is It Really Mobility's Great Hope?" *CityLab*, October 25, 2018, https://www.citylab.com/perspective/2018/10/helsinkis-maas-app-whim-is-it-really-mobilitys-great-hope/573841/.

111 **the service will offer side-by-side comparison**: Andrew J. Hawkins, "Uber Just Added Public Transportation to Its App," *The Verge*, January 31, 2019, https://www.theverge.com/2019/1/31/18205154/uber-public-transportation-app-denver.

5. Continuous Delivery

115 **"For e-commerce firms"**: Liu Yi, "Can Jack Ma's Cainiao Project Deliver on Its Promise?" *South China Morning Post*, July 13, 2013, https://www.scmp.com/news/china/article/1281450/can-jack-mas-cainiao-project-deliver-its-promise.

116 **average size of newly built houses in the US**: US Census Bureau and US Department of Housing and Urban Development, *2014 Characteristics of New Housing* (2015), 345, https://www.census.gov/construction/chars/pdf/c25ann2014.pdf.

116 **"The home transformed into a warehouse"**: Alex Evans, "Retail, Owner-ship and Deflation in the Last Mile," *Hackernoon*, October 26, 2017, https://hackernoon.com/retail-ownership-and-winning-the-last-mile-86ef4eb1a7e7.

116 **closer to 250**: Federal Highway Administration, *Trends in Discretionary Travel: 2017 National Household Travel Survey*, FHWA NHTS Report (Washington, DC: FHWA, February 2019).

116 **the number could fall another 30 percent**: KPMG, *Autonomy Delivers: An Oncoming Revolution in the Movement of Goods* (white paper, KPMG, 2018), 11.

116 **free or discounted shipping was the main reason**: UPS, *Pulse of the Online Shopper Study* (April 2018).

117 **"Driving to shop seems convenient"**: KPMG, *Autonomy Delivers*, 3.

117 **online retailers took in more than half a trillion**: US Census Bureau, "Quarterly Retail E-commerce Sales: 1st Quarter 2019," *US Census Bureau News*, CB19-63.

117 **3 percent of groceries in America**: Alana Semuels, "Why People Still Don't Buy Groceries Online," *The Atlantic*, February 5, 2019, https://www.theatlantic.com/technology/archive/2019/02/online-grocery-shopping-has-been-slow-catch/581911/.

117 **up to 35 percent of all consumer spending in the UK**: David Jinks Milt, *2030: The Death of the High Street*, ParcelHero Industry Report, 2, https://www.parcelhero.com/content/downloads/pdfs/high-street/deathofthehighstreetreport.pdf; **and China**: "2019: China to Surpass US in Total Retail Sales," *eMarketer*, January 23, 2019, https://www.emarketer.com/newsroom/index.php/2019-china-to-surpass-us-in-total-retail-sales/.

117 **another 12,000 would shutter**: Glenn Taylor, "Report: Store Closures to Reach 12,000 in 2018," *Retail TouchPoints*, January 2, 2018, https://www.retailtouchpoints.com/features/news-briefs/report-store-closures-to-reach-12-000-in-2018.

117 **as many announced store closures as in the entire preceding year**: Sapna Maheshwari, "US Retail Stores' Planned Closings Already Exceed 2018 Total," *New York Times*, April 12, 2019, https://www.nytimes.com/2019/04/12/business/retail-store-closings.html.

117 **one-quarter of shopping malls in the US will close**: "Apparel Retail and Brands: Making Sense of Softlines following a Tumultuous Twelve Months," Credit Suisse Research, May 2017, https://research-doc.credit-suisse.com/docView?document_id=1075851631.

118 **$141 billion in product sales in 2018**: Allison Enright, "Amazon's Prod-

uct Sales Climb Nearly 20% in 2018, but Only 8% in Q4," *Digital Commerce 360*, January 31, 2019, https://www.digitalcommerce360.com/article/amazon-sales/.

118 **52 cents of every dollar spent online**: Lauren Thomas, "74% of Consumers Go to Amazon When They're Ready to Buy Something. That Should Be Keeping Retailers Up at Night," *CNBC*, March 19, 2019, https://www.cnbc.com/2019/03/19/heres-why-retailers-should-be-scared-of-amazon-dominating-e-commerce.html.

118 **its share is growing**: "Top 10 US Companies, Ranked by Retail Ecommerce Sales Share, 2018 (% of US Retail Ecommerce Sales)," *eMarketer*, July 17, 2018, https://www.emarketer.com/Chart/Top-10-US-Companies-Ranked-by-Retail-Ecommerce-Sales-Share-2018-of-US-retail-ecommerce-sales/220521.

118 **Amazon is worth more than three times**: Erica Pandey, "On the Mind of Every Retail CEO: The Amazon Threat," *Axios*, September 19, 2018, https://www.axios.com/amazon-strategy-shopify-flipkart-macys-code-commerce-1e1a5fae-75b0-4a68-b308-9ca1a539437d.html.

118 **Alibaba has a similar market share**: "2019: China to Surpass US."

118 **half of all parcel shipments in the US**: Tim Laseter, Andrew Tipping, and Fred Duiven, "The Rise of the Last-Mile Exchange," *Strategy + Business*, July 30, 2018, https://www.strategy-business.com/article/The-Rise-of-the-Last-Mile-Exchange.

119 **Alibaba delivered *one billion* shipments**: Lisa Lacy, "Alibaba Rings Up $30.8 Billion on Singles Day 2018," *Adweek*, November 11, 2018, https://www.adweek.com/digital/alibaba-rings-up-30-8-billion-on-singles-day-2018/.

119 **million-person courier team**: Rita Liao, "Alibaba and Amazon Move Over, We Visited JD's Connected Grocery Store in China," *TechCrunch*, November 15, 2018, https://techcrunch.com/2018/11/15/jd-7fresh-supermarket/.

119 **spent more than $20 *billion* on shipping**: Alana Semuels, "Free Shipping Isn't Hurting Amazon," *The Atlantic*, April 27, 2018, https://www.theatlantic.com/technology/archive/2018/04/free-shipping-isnt-hurting-amazon/559052/.

119 **10 percent of its total take**: Daphne Howland, "Amazon Captured 44% of US Online Sales Last Year," *Retail Dive*, January 4, 2018, https://www.retaildive.com/news/amazon-captured-44-of-us-online-sales-last-year/514044/.

119 **same-day delivery is still very limited**: Julian Allen, Maja Piecyk, and Marzena Piotrowska, *An Analysis of the Same-Day Delivery Market and Operations in the UK*, Technical Report CUED/C-SRF/TR012 (Westminster, UK: University of Westminster, November 2018), 10.

119 **consumers will expect *all* deliveries**: McKinsey & Company, *An Integrated*

Perspective on the Future of Mobility (McKinsey Center for Future Mobility, October 2016), 3.

120 **"too much change in too short a period of time"**: Alvin Toffler, *Future Shock* (New York: Random House, 1970).

120 **receives five deliveries each week, twice as many**: N. McGuckin and A. Fucci, *Summary of Travel Trends: 2017 National Household Travel Survey*, Report No. FHWA-PL-18-019 (Washington, DC: Federal Highway Administration, July 2018), 100.

121 **schlepp stuff into the surrounding territory**: Joann Muller, "One Big Thing: The Rise of Driverless Delivery," *Axios*, November 28, 2018, https://www.axios.com/autonomous-vehicles-could-be-used-for-deliveries-3fb12a24-3e66-4d8b-b678-a2fbb47d05cb.html.

121 **By 2020 fully two-thirds will be**: Laseter et al., "The Rise of the Last-Mile Exchange."

122 **more than *twice* as many shipments**: Laseter et al., "The Rise of the Last-Mile Exchange."

122 **exhausted racers mustered what energy they could**: Dan Peterson, "Why Are Marathons 26.2 Miles Long?" *Live Science*, April 19, 2010, https://www.livescience.com/11011-marathons-26-2-miles-long.html.

123 **a thankless, never-ending battle against backhoes**: Andrew Blum, *Tubes: A Journey to the Center of the Internet* (New York: Ecco, 2012), 1–10.

123 **800,000 square feet of order-fulfillment space**: Allen et al., *An Analysis of the Same-Day Delivery*, 76.

123 **"Large shipments of goods atomize into hundreds"**: Ted Choe et al., "The Future of Freight: How New Technology and New Thinking Can Transform How Goods Are Moved," *Insights*, Deloitte, June 28, 2017, https://www2.deloitte.com/insights/us/en/focus/future-of-mobility/future-of-freight-simplifying-last-mile-logistics.html.

123 **Amazon's newest fulfillment centers**: Janelle Jones and Ben Zipperer, "Unfulfilled Promises: Amazon Fulfillment Centers Do Not Generate Broad-Based Employment Growth," Economic Policy Institute, February 1, 2018, https://www.epi.org/publication/unfulfilled-promises-amazon-warehouses-do-not-generate-broad-based-employment-growth/.

123 **In big cities now you can find them in a circle**: Patrick Kiger, "Driving Hard to Secure Last-Mile Logistics," Urban Land Institute, February 5, 2018, https://urbanland.uli.org/industry-sectors/industrial/driving%E2%80%85hard-secure-last-mile-logistics/.

124 **a veteran local driver**: Allen et al., *An Analysis of the Same-Day Delivery*, 141.

124 **deskilling of local delivery has driven down wages**: Anoosh Chake-lian, "'Slaveroo': How Riders Are Standing Up to Uber, Deliveroo and the Gig Economy," *New Statesman America*, September 24, 2018, https://www.newstatesman.com/politics/uk/2018/09/slaveroo-how-riders-are-standing-uber-deliveroo-and-gig-economy.

125 **slash the cost of delivery by 90 to 95 percent**: "Starship Technologies Inc.," Robotics Business Review, accessed February 26, 2019, https://www.roboticsbusinessreview.com/listing/starship-technologies/.

125 **deskside delivery allowed workers to skip**: "Starship Technologies Launches Commercial Rollout of Autonomous Delivery," Starship, April 30, 2018, https://www.starship.xyz/press_releases/2708/.

125 **replacing a swarm of dirty, noisy, and often dangerous delivery trucks**: "The Future of Moving Things," IDEO, accessed June 11, 2018, https://automobility.ideo.com/moving-things/a-new-familiar-sight.

125 **raised $940 million in venture funding**: Mary Ann Azevedo, "Nuro Raises $940M from SoftBank Vision Fund for Robot Delivery," *Crunchbase*, February 11, 2019, https://news.crunchbase.com/news/nuro-raises-940m-from-softbank-vision-fund-for-robot-delivery/.

125 **Toyota's mule, the e-Palette**: Andrew J. Hawkins, "Toyota's 'e-Palette' Is a Weird, Self-Driving Modular Store on Wheels," *The Verge*, January 8, 2018, https://www.theverge.com/2018/1/8/16863092/toyota-e-palette-self-driving-car-ev-ces-2018.

125 **The Puppy 1, made in China, even balances**: Greg Nichols, "Behold, the Self-Driving Suitcase," *ZDNet*, February 7, 2018, https://www.zdnet.com/article/behold-the-self-driving-suitcase/.

125 **a future for its Sprinter delivery vans**: "Mercedes-Benz Vans Invests in Starship Technologies, the World's Leading Manufacturer of Delivery Robots," Daimler, January 13, 2017, https://media.daimler.com/marsMediaSite/ko/en/15274799.

126 **Roadie, for instance, brokers shipments of bulky, slightly**: Adele Peters, "This Package Rideshare App Pays to Use Your Empty Trunk," *Fast Company*, February 5, 2015, https://www.fastcompany.com/3041884/this-package-rideshare-app-pays-to-use-your-empty-trunk.

127 **claims to have a footprint larger than Amazon Prime**: "Roadie: About Us," Roadie, accessed June 12, 2018, https://www.roadie.com/about.

127 **a cheaper, better, and bigger alternative**: "Lockers: Deliveries and Returns Made Easy," Amazon, accessed June 12, 2018, https://www.amazon.com/b?ie=UTF8&node=6442600011.

128 **25,000 "milk floats," tiny delivery carts with open sides**: Bonnie Alter, "Electric Milk Trucks Still Working in Jolly Old England," *TreeHugger*, April 12, 2012, https://www.treehugger.com/cars/electric-milk-trucks-still-working -jolly-old-england.html.

128 **"Short ranges and low top speed"**: "Opportunities Are Opening for Elec- trified Commercial Vehicles," *The Economist*, February 15, 2018, https:// www.economist.com/business/2018/02/15/opportunities-are-opening-for -electrified-commercial-vehicles.

128 **bulking up reduces per-unit delivery costs significantly**: McKinsey & Company, *An Integrated Perspective*, 4.

129 **more lucrative to rent night mules out as energy storage**: Henning Heppner, interview with author, November 9, 2018.

130 **"There is little reason to doubt"**: Edward Glaeser and Janet Kohlhase, "Cit- ies, Regions, and the Decline of Transport Costs," *Papers in Regional Science* 83 (2004): 197–228.

130 **each promise huge cost savings**: McKinsey & Company, *An Integrated Per- spective*, 22.

132 **"They're not waiting for a package"**: National League of Cities, "Jobs and Economy: A Human Touch for Robot Delivery," *Autonomous Vehicles: Future Scenarios*, accessed February 26, 2019, http://avfutures.nlc.org/jobs-and -the-economy.

132 **"Amazon Day" promotion**: Taylor Nicole Rogers, "Amazon's New Waste- Reduction Strategy: Deliver Only Once a Week," *CNN*, February 28, 2019, https://www.cnn.com/2019/02/28/tech/amazon-day/index.html.

133 **most often impatient, high-spending young shoppers**: McKinsey & Com- pany, *Parcel Delivery: The Future of Last Mile* (September 2016), 6.

6. Creative Destruction

135 **"What we're trying to do at Pizza Hut"**: Tim Higgins, "Pizza Delivery Gears Up for a Driverless Era," *Wall Street Journal*, June 26, 2018, https://www.wsj .com/articles/pizza-delivery-may-be-entering-a-new-era-1530029087.

136 **1.5 million packages each day, triple the number a decade ago**: Mat- thew Haag and Winnie Hu, "1.5 Million Packages a Day: The Internet Brings Chaos to N.Y. Streets," *New York Times*, October 28, 2019, https://www.nytimes .com/2019/10/27/nyregion/nyc-amazon-delivery.html.

136 **Amazon cornered the market**: Scott Kirsner, "Acquisition Puts Amazon Rivals in Awkward Spot," *Boston Globe*, December 1, 2013, https://www.bos-

tonglobe.com/business/2013/12/01/will-amazon-owned-robot-maker-sell
-tailer-rivals/FON7bVNKvfzS2sHnBHzfLM/story.html.

136 **100,000 of these diligent devices were reporting:** Nick Wingfield, "As Amazon Pushes Forward with Robots, Workers Find New Roles," *New York Times*, September 10, 2017, https://www.nytimes.com/2017/09/10/technology/amazon robots-workers.html.

137 **cramming 50 percent more inventory:** Ananya Bhattacharya, "Amazon Is Just Beginning to Use Robots in Its Warehouses and They're Already Making a Huge Difference," *Quartz*, June 17, 2016, https://qz.com/709541/amazon -is-just-beginning-to-use-robots-in-its-warehouses-and-theyre-already -making-a-huge-difference/.

137 **bots that slide along tracks suspended:** Fiona Hartley, "Over 1,000 Robots Pack Groceries in Ocado's Online Shopping Warehouse," *Dezeen*, June 6, 2018, https://www.dezeen.com/2018/06/06/video-ocado-warehouse-shopping -robots-movie/.

137 **China's JD.com employs a freakish spiderlike robot:** JD.com, "Tour of the Warehouse of the Future," YouTube video, May 11, 2017, https://www.youtube .com/watch?v=24T14VP5vVs.

137 **will face excruciating pressure to control carbon emissions:** OECD/ITF, *ITF Transport Outlook 2017* (Paris: OECD Publishing, 2017), https://www.ttm .nl/wp-content/uploads/2017/01/itf_study.pdf.

137 **"that incessantly revolutionizes the economic structure":** Joseph Schumpeter, *Capitalism, Socialism, and Democracy* (London, UK: Routledge, 1942), 82–83.

138 **Domino's Pizza outlet averages a mere 1,500 square feet:** United States Securities and Exchange Commission, *Annual Report Pursuant to Section 13 or 15(d): Domino's Pizza*, December 29, 2013, p. 4, https://www.sec.gov/Archives/ edgar/data/1286681/000119312514066092/d661353d10k.htm.

138 **cardboard pizza, scant sauce, and synthetic-tasting:** Allison P. Davis, "How Domino's Became the Pizza for the People," *The Ringer*, February 28, 2017, https://www.theringer.com/2017/2/28/16043242/dominos-pizza -tracker-pan-recipe-superfans-db39239ed88e.

139 **packed nine ghost restaurants:** Robert Channick, "9 Restaurants, 1 Kitchen, No Dining Room—Virtual Restaurants Open for Online Delivery," *Chicago Tribune*, March 27, 2017, https://www.chicagotribune.com/business/ct-chicago -virtual-restaurants-0328-biz-20170309-story.html.

139 **plans to open 400 locations:** Joseph Pimentel, "Novelty or Next Trend? Google Venture-Backed Kitchen United Looks to Open 400 Ghost Kitchens

Nationwide," Bisnow Los Angeles, July 21, 2019, https://www.bisnow.com/los
-angeles/news/retail/novelty-or-next-trend-google-backed-kitchen-united
-looks-to-open-400-virtual-food-halls-nationwide-100026.

139 **Green Summit needs just one-quarter of the floor space**: Channick, "9 restaurants, 1 kitchen."

140 **Traditional restaurants could easily spend 10 times**: Neil Ungerleider,
"Hold the Storefront: How Delivery-Only 'Ghost' Restaurants Are Changing Takeout," *Fast Company*, January 20, 2017, https://www.fastcompany
.com/3064075/hold-the-storefront-how-delivery-only-ghost-restaurants-are
-changing-take-out.

140 **more than $10 billion flowed into last-mile**: McKinsey & Company, *Parcel
Delivery*, 6.

140 **group took control of four in New York City**: Mary Diduch, "Travis Kalanick
Said Last Year He Was Getting into Real Estate. Here's What He's Buying in New
York," *The Real Deal*, February 4, 2019, https://therealdeal.com/2019/02/04/
travis-kalanick-said-last-year-he-was-getting-into-real-estate-heres-what
-hes-buying-in-nyc/.

140 **a go-to spot for same-day distribution**: Keiko Morris, "Online Commerce
Sparks Industrial Real-Estate Boom off the Beaten Path in N.J.," *Wall Street
Journal*, February 26, https://www.wsj.com/articles/online-commerce
-sparks-industrial-real-estate-boom-off-the-beaten-path-in-n-j-1488150878.

141 **nation's largest dry-cleaning operation**: Rebecca Greenfield, "Inside Rent
the Runway Secret Dry-Cleaning Empire," *Fast Company*, October 28, 2014,
https://www.fastcompany.com/3036876/inside-rent-the-runways-secret
-dry-cleaning-empire.

141 **dwarfed only by the company's second plant**: "Fashion Company, Rent
the Runway, to Open Arlington Distribution Center," Arlington, February 28,
2018, http://arlington.hosted.civiclive.com/news/my_arlington_t_x/news_
archive/2018_archived_news/february_2018/fashion_company_rent_the_
runway_to_open.

142 **Domino's teamed up with mule manufacturer Nuro**: "Nuro and Domino's
Partner on Autonomous Pizza Delivery," *Nuro* (blog), Medium, June 17, 2019,
https://medium.com/nuro/nuro-and-dominos-partner-on-autonomous
-pizza-delivery-88c6b6640ff0.

142 **AVs carrying an entire kitchen**: "Toyota Launches New Mobility Ecosystem
and Concept Vehicle at 2018 CES," Toyota, January 9, 2018, https://newsroom
.toyota.co.jp/en/corporate/20546438.html.

142 **increase in curbside cardboard collection**: Nicole Nguyen, "The Hid-

den Environmental Cost of Amazon Prime's Free, Fast Shipping," *BuzzFeed News*, July 21, 2018, https://www.buzzfeednews.com/article/nicolenguyen/environmental-impact-of-amazon-prime.

142 **"Kipple"**: Philip K. Dick, *Do Androids Dream of Electric Sheep?* (New York: Random House, 1968), 65.

143 **"When nobody's around, kipple reproduces itself"**: Dick, *Do Androids Dream*, 65.

145 **The rebound effect shows up everywhere**: David Owen, "The Efficiency Dilemma," *New Yorker*, December 20 & 27, 2010, https://www.newyorker.com/magazine/2010/12/20/the-efficiency-dilemma.

145 **eating more pizza than ever**: Mona Chalabi, "A National Pizza Day Investigation: How Many Slices a Day Do Americans Eat?" *The Guardian*, February 9, 2017, https://www.theguardian.com/lifeandstyle/datablog/2017/feb/09/national-pizza-day-how-many-slices-do-americans-eat.

146 **a design movement than a solid theory**: "What Is a Circular Economy?" Ellen MacArthur Foundation, accessed March 6, 2019, https://www.ellenmacarthurfoundation.org/circular-economy/concept.

147 **AVs will make more than 300,000 instant deliveries**: McKinsey & Company, *Parcel Delivery*, 18.

147 **the meal's million-mile journey**: "Supermarkets in China: Two Ma Race," *The Economist*, April 7, 2018, 55.

147 **Lufa Farms provides personalized, online ordering**: Sarah Treleaven, "Is Personalized, Next-Day Delivery the Future of Urban Farming?" *CityLab*, February 9, 2018. https://www.citylab.com/environment/2018/02/is-personalized-next-day-delivery-the-future-of-urban-farming/551981/.

148 **8 percent of global carbon emissions**: Elizabeth Cline, *The Conscious Closet: The Revolutionary Guide to Looking Good While Doing Good* (New York: Plume, 2019), 3.

148 **textiles make up 5 percent of the waste**: Mireya Navarro, "Don't Toss That Old Shirt. They'll Pick It Up," *New York Times*, May 25, 2010, https://www.nytimes.com/2010/05/26/nyregion/26clothing.html.

148 **company has grown to serve some 120,000 customers**: Mizuho Aoki, "Japan's Budding Fashion Rental Services Proving Popular with Working Women, Moms," *Japan Times*, August 18, 2017, https://www.japantimes.co.jp/news/2017/08/18/business/japans-budding-fashion-rental-services-proving-popular-working-women-moms.

149 **25 to 40 percent of the apparel ordered online is returned**: Steve Dennis, "Many Unhappy Returns: E-commerce's Achilles Heel," *Forbes*, August 9, 2017,

https://www.forbes.com/sites/stevendennis/2017/08/09/many-unhappy
-returns-e-commerces-achilles-heel/#67b40f764bf2.

149 **"If anything, it feels more like a warning"**: Melanie Ehrenkranz, "This Game
Imagines a Bleak Future in Which Tech Companies Win and Gig Workers Just
Try to Survive," *Gizmodo*, March 23, 2018, https://gizmodo.com/this-game
-imagines-a-bleak-future-in-which-tech-compani-1824021429.

150 **the trio worked out an ingenious and comprehensive new approach**:
David H. Autor, Frank Levy, and Richard J. Murnane, "The Skill Content of
Recent Technological Change: An Empirical Exploration," *Quarterly Journal
of Economics* (2003): 1279–333.

150 **"they can be exhaustively specified"**: Autor et al., "The Skill Content," 1283.

152 **predictions have also proved remarkably prescient**: The *Quarterly Journal
of Economics* article has been cited more than 4,300 times in the last 15 years,
according to Google Scholar.

152 **"limited opportunities for substitution or complementarity"**: Chris
O'Brien, "Venture Capitalists Flock to Truck Technology Startups,"
Trucks.com, July 31, 2017, https://www.trucks.com/2017/07/31/venture
-capitalists-flock-truck-technology-startups/.

152 **an "informal" attempt to chart the territory**: Autor et al., "The Skill Content," 1282.

153 **jobs in the US economy were at risk of automation**: Carl Benedikt Frey and
Michael A. Osborne, "The Future of Employment: How Susceptible Are Jobs to
Computerisation?" *Technological Forecasting and Social Change* 114 (January
2017): 254–80, https://www.oxfordmartin.ox.ac.uk/downloads/academic/
The_Future_of_Employment.pdf.

154 **SAFE has been a strong advocate of AVs**: James Osborne, "Among Retired
Military Leaders, U.S. Thirst for Oil Poses Security Risk," *Houston Chronicle*,
July 8, 2017, https://www.houstonchronicle.com/business/article/Among
-retired-military-leaders-U-S-thirst-for-11272912.php.

154 **economists see professional drivers in the crosshairs of automation's
advance**: Securing America's Future Energy, *America's Workforce and the
Self-Driving Future: Realizing Productivity Gains and Spurring Economic
Growth*, June 2018, 8, https://avworkforce.secureenergy.org/wp-content/
uploads/2018/06/SAFE_AV_Policy_Brief.pdf.

155 **benefits vastly outweighs projected wage losses**: Securing America's
Future Energy, *America's Workforce*, 12.

155 **factor of three or more**: Securing America's Future Energy, *America's
Workforce*, 9.

155 **plenty of time for policymakers to prepare**: Securing America's Future Energy, *America's Workforce*, 8.

155 **three possible approaches for the US**: Securing America's Future Energy, *America's Workforce*, 44.

156 **as the trucking workforce ages**: David H. Freedman, "Self-Driving Trucks," *MIT Technology Review*, March/April 2017, https://www.technologyreview .com/s/603493/10-breakthrough-technologies-2017-self-driving-trucks/.

156 **400,000 seasoned drivers to retire**: "The Future of Trucking: Mixed Fleets, Transfer Hubs, and More Opportunity for Truck Drivers," *Uber* (blog), Medium, February 1, 2018. https://medium.com/@UberATG/the-future-of-trucking -b3d2ea0d2db9.

156 **involved in accidents at twice the rate**: David Cullen, "Why Insurance Costs Are Sky High," *Truckinginfo.com*, October 19, 2017, https://www.truckinginfo .com/157754/why-insurance-costs-are-sky-high.

156 **pay-per-mile insurance services**: Sheffi Ben-Hutta, "HUMN.ai Completes Successful London Fleet Insurance Pilot," *Coverager*, November 16, 2018, https://www.coverager.com/humn-ai-completes-successful-london-fleet -insurance-pilot/.

157 **automated AVs could cut the cross-country**: "Could Self-Driving Trucks Make Roads Safer?" *EHS Today*, February 20, 2018, https://www.ehstoday .com/safety-technology/could-self-driving-trucks-make-roads-safer.

157 **Young drivers are more inclined**: Julian Allen, Maja Piecyk, and Marzena Piotrowska, *An Analysis of the Same-Day Delivery Market and Operations in the UK*, Technical Report CUED/C-SRF/TR012 (Westminster, UK: University of Westminster, November 2018), 141.

157 **trimmed billions of dollars of inventory**: "Amazon and Alibaba Are Pacesetters of the Next Supply-Chain Revolution," *The Economist*, July 11, 2019, https:// www.economist.com/special-report/2019/07/11/amazon-and-alibaba-are -pacesetters-of-the-next-supply-chain-revolution.

158 **forced to cut holes in the structure's roof**: Katie Linsell and Ellen Milligan, "Robot Workers Can't Go on Strike but They Can Go Up in Flames," *Bloomberg*, March 3, 2019, https://www.bloomberg.com/news/articles/2019-03-03/robot -workers-can-t-go-on-strike-but-they-can-go-up-in-flames.

7. The New Highwaymen

161 **"The purpose of a system is what it does"**: Stafford Beer, "What Is Cybernetics?" *Kybernetes* 31, no. 2 (2002), doi:10.1108/03684920210417283.

161 **the last mounted heist in England**: "Highwaymen of the Peak," *BBC*, July 7, 2003, http://www.bbc.co.uk/insideout/eastmidlands/series3/travellers_ highwaymen_derbyshire_peakdistrict.shtml.

161 **One popular punishment was gibbeting**: "Highwaymen: Capture and Punishment," *The Gazette*, accessed March 14, 2019, https://www.thegazette.co.uk/all-notices/content/100465.

162 **organizations issued bonds and levied tolls to finance them**: *Abstract of General Statements of Income and Expenditure of Turnpike Trusts in England and Wales, from 1st January 1838 to 31st December 1838, inclusive*, 1838, 3 & 4 Will IV, chapter 80.

162 **turnpike trusts boosted property values**: Dan Bogart, "The Turnpike Roads of England and Wales," Cambridge Group for the History of Population and Social Structure (University of Cambridge), 1, accessed March 13, 2019, https://www.campop.geog.cam.ac.uk/research/projects/transport/onlineatlas/britishturnpiketrusts.pdf.

162 **"meant they were basically able to determine"**: Philip Thomas, "The Rebecca Riots by Local Historian Philip Thomas," *West Wales Chronicle*, October 10, 2018, https://www.westwaleschronicle.co.uk/blog/2018/10/10/the-rebecca-riots-by-local-historian-columnist-philip-thomas/.

162 **"And they blessed Rebekah and said unto her"**: Gen. 24:60.

163 **prompting the gang of crossdressers to sally forth**: Thomas, "The Rebecca Riots."

163 **financial intermediaries reached an all-time high**: Alan S. Binder et al., *Rethinking the Financial Crisis* (New York: Russel Sage Foundation, 2012).

164 **unknown in America before World War II, ballooned**: Derek Fidler and Hicham Sabir, "The Cost of Housing Is Tearing Our Society Apart," World Economic Forum, part of the World Economic Forum Annual Meeting, January 9, 2019, https://www.weforum.org/agenda/2019/01/why-housing-appreciation-is-killing-housing/.

164 **half the price of a barrel of oil**: Tracy Alloway, "Dresdner/Commerzbank Blames Oil Speculators," *Financial Times*, August 21, 2009, https://ftalphaville.ft.com/2009/08/21/68101/dresdnercommerzbank-blames-oil-speculators/.

164 **Most of the dead were farmers**: Fredrick Kaufman, *Bet the Farm: How Food Stopped Being Food* (New York: John Wiley & Sons, 2012).

165 **"uninterested in material comfort that he barely knew"**: Janny Scott, "After 3 Days in the Spotlight, Nobel Prize Winner Is Dead," *New York Times*, October 12, 1996, https://www.nytimes.com/1996/10/12/nyregion/after-3-days-in-the-spotlight-nobel-prize-winner-is-dead.html.

165 **the transit authority's financial outlook was grim**: Ari L. Goldman, "Ridership of Subway since 1917," *New York Times*, October 23, 1982, https://www.nytimes.com/1982/10/23/nyregion/ridership-of-subways-since-1917.html.

166 **marginal costs of rush-hour service into Manhattan's**: William S. Vickrey, *The Revision of the Rapid Transit Fare Structure of the City of New York*, Technical Monograph No. 3 (New York: Mayor's Committee on Management Survey of the City of New York, 1952), 24–53, https://babel.hathitrust.org/cgi/pt?id=mdp.39015020930130.

166 **a refundable deposit, to be collected**: Vickrey, *The Revision of the Rapid Transit Fare Structure*, 98.

167 **"Transit pricing still appears to be set"**: Richard Arnott, "William Vickrey: Contributions to Public Policy," Department of Economics, Boston College, October 1997, http://fmwww.bc.edu/ec-p/wp387.pdf.

167 **Singapore's Area Licensing Scheme**: Scott, "After 3 Days in the Spotlight."

167 **using carpools and buses expanded**: Kenneth A. Small and Jose Gomez-Ibanez, "Road Pricing for Congestion Management: The Transition from Theory to Policy," in *Road Pricing, Traffic Congestion, and the Environment*, ed. Kenneth J. Button and Eric T. Verhoef (Cheltenham, UK: Edward Elgar, 1998), 216.

167 **London (in 2003)**: Transport for London, *Congestion Charging Central London Impacts Monitoring Second Annual Report* (London, UK: Transport for London, April 2004), 1–6; **Stockholm (in 2006)**: Tri-State Transportation Campaign, *Road Pricing in London, Stockholm, and Singapore* (New York: Tri-State Transportation Campaign, 2018), 14–17.

168 **least likely to own cars**: Transit Center, *Subsidizing Congestion: The Multibillion-Dollar Tax Subsidy That's Making Your Commute Worse* (New York: Transit Center and Frontier Group, 2018), 2–5; **they gain the most**: "Congestion Pricing Would Save Riders of Most Queens and Brooklyn Express Buses One to Two Hours per Week," *Riders Alliance* (blog), Medium, October 30, 2018, https://medium.com/@RidersNY/congestion-pricing-would-save-queens-brooklyn-express-bus-riders-1-to-2-hours-per-week-ed967dfbfdc0.

168 **a whopping 73 percent of the increased traffic**: San Francisco County Transportation Authority, *TNCs and Congestion Draft Report* (San Francisco, CA: SFCTA, October 2018), 28.

169 **the *cisia***: "The Odometer by Vitruvius and Heron," *YourForum*, May 3, 2009, http://yourforum.gr/InvisionBoard/The-Oedoemeter-By-Vitrueviues-Aend-Heroen-t174792.html.

169 **a parts budget of only three dollars**: Ronald Harstad, "William S. Vickrey," Working Papers 0519, Department of Economics, University of Missouri, 2005, 3.

172 **how airline tickets are still sold**: William S. Vickrey, "Some Objections to Marginal Cost Pricing," *Journal of Political Economy* 56 (1948): 218–38, https://www.journals.uchicago.edu/doi/abs/10.1086/256674?journalCode=jpe.

172 **to bring in more than $1 billion**: Elise Young and Henry Goldman, "Congestion Pricing. Has It Worked and Can It Fix New York City?" *Washington Post*, April 1, 2019, https://www.washingtonpost.com/business/congestion-pricing-has-it-worked-and-can-it-fix-new-york/2019/04/01/c8b360d2-5493-11e9-aa83-504f086bf5d6_story.html.

173 **"Since then, rates have skyrocketed"**: Aaroncynic, "Chicago's Awful Parking Meters Make Big Bucks for Private Investors Again," *Chicagoist*, May 24, 2016, https://chicagoist.com/2016/05/24/parking_meters_make_big_bucks_for_p.php.

174 **"horrible monster, whose tentacles spread poverty"**: Eric Posner and Glen Weyl, "The Real Villain behind Our New Gilded Age," *New York Times*, May 1, 2018, https://www.nytimes.com/2018/05/01/opinion/monopoly-power-new-gilded-age.html.

174 **"public transportation in large American cities"**: Charles W. Cheape, *Moving the Masses: Urban Public Transit in New York, Boston, and Philadelphia 1880–1912* (Cambridge, MA: Harvard University Press, 1980), 1.

175 **companies merely joined forces**: Cheape, *Moving the Masses*, 172.

175 **the most powerful, reviled traction monopoly**: Walt Crowley, "City Light's Birth and Seattle's Early Power Struggles, 1886–1950," History Link, April 26, 2000, https://www.historylink.org/File/2318.

175 **enjoyed decades of unrivaled power**: Owain James, "We Miss Streetcars' Frequent and Reliable Service, Not Streetcars Themselves," *Mobility Lab*, April 17, 2019, https://mobilitylab.org/2019/04/17/we-miss-streetcars-frequent-and-reliable-service-not-streetcars-themselves/; **combination of technological change and federal intervention:** "Jersey Trolley Merger," *Wall Street Journal*, May 13, 1905, 2.

176 **$100 billion Vision Fund**: Katrina Brooker, "The Most Powerful Person in Silicon Valley," *Fast Company*, January 14, 2019, https://www.fastcompany.com/90285552/the-most-powerful-person-in-silicon-valley.

176 **its total commitment to some $9 billion**: Pavel Alpeyev, Jie Ma, and Won Jae Ko, "Taxi-Hailing Apps Take Root in Japan as SoftBank, Didi Join Fray,"

Bloomberg, July 19, 2018, https://www.bloomberg.com/news/articles/2018-07
-19/softbank-didi-to-roll-out-taxi-hailing-business-in-japan.

177 **$2 billion into Singapore-based Grab**: Yoolim Lee, "Grab Vanquishes Uber
with Local Strategy, Billions from SoftBank," *Bloomberg*, March 26, 2018,
https://www.bloomberg.com/news/articles/2018-03-26/grab-vanquishes
-uber-with-local-strategy-billions-from-softbank.

177 **Ola downloaded $2 billion**: Saritha Rai, "India's Ola Raises $2 Billion from
SoftBank, Tencent," *Bloomberg*, October 2, 2017, https://www.bloomberg
.com/news/articles/2017-10-02/india-s-ola-is-said-to-raise-2-billion-from
-softbank-tencent.

177 **15 percent stake in Uber**: Alison Griswold, "SoftBank—not Uber—Is the
Real King of Ride-Hailing," *Quartz*, January 23, 2018, https://qz.com/1187144/
softbank-not-uber-is-the-real-king-of-ride-hailing/.

177 **Uber picked off Dubai-based Careem**: Adam Satariano, "This Estonian
Start-Up Has Become a Thorn in Uber's Side," *New York Times*, April 23, 2019,
https://www.nytimes.com/2019/04/23/technology/bolt-taxify-uber-lyft.html.

177 **The damage to consumers**: Justina Lee, "Singapore Fine Is 'Minor Bump'
in Grab's Ride-Hailing Dominance," *Nikkei Asian Review*, September 25,
2018, https://asia.nikkei.com/Spotlight/Sharing-Economy/Singapore-fine-is
-minor-bump-in-Grab-s-ride-hailing-dominance.

177 **Grab cornered more than 80 percent**: Ardhana Aravindan, "Singapore
Fines Grab and Uber, Imposes Measures to Open Up Market," *Reuters*, September 23, 2018, https://www.reuters.com/article/us-uber-grab-singapore/
singapore-fines-grab-and-uber-imposes-measures-to-open-up-market
-idUSKCN1M406J.

177 **all launched antitrust investigations**: Mai Nguyen, "Vietnam Says Eyeing
Formal Antitrust Probe into Uber-Grab Deal," *Reuters*, May 16, 2018, https://
www.reuters.com/article/us-uber-grab-vietnam-idUSKCN1IH0XN; Aika Rey,
"Antitrust Watchdog Fines Grab P16 Million over Uber Deal," *Rappler*, October
17, 2018, https://www.rappler.com/business/214502-philippine-competition
-commission-fines-grab-philippines-over-uber-deal; Yoolim Lee, "Singapore
Watchdog Fines Uber, Grab $9.5 Million over Merger," *Bloomberg*, September
24, 2018, https://www.bloomberg.com/news/articles/2018-09-24/singapore
-fines-uber-grab-s-13-million-for-merger-infringement.

177 **another fare-slashing battle with Ola**: "Steering Group: A Bold Scheme to
Dominate Ride-Hailing," *The Economist*, May 10, 2018, https://www.economist
.com/briefing/2018/05/10/a-bold-scheme-to-dominate-ride-hailing.

177 **"SoftBank is playing the ride-hailing"**: Alison Griswold, "Softbank Has Spread Its Ride-Hailing Bets and Didi Looks Like an Early Win," *Quartz*, April 24, 2018, https://qz.com/1261177/softbanks-winner-in-ride-hailing-is-chinas -didi-chuxing-not-uber/.

177 **"driver incentives, passenger discounts"**: Tim O'Reilly, "The Fundamental Problem with Silicon Valley's Favorite Growth Strategy," *Quartz*, February 5, 2019, https://qz.com/1540608/the-problem-with-silicon-valleys-obsession -with-blitzscaling-growth/.

178 **"locked in a capital-fueled deathmatch"**: O'Reilly, "The Fundamental Problem."

178 **The Vision Fund's biggest investor**: Brooker, "The Most Powerful Person."

178 **the proceeds of an earlier liquidation**: Catherine Shu, "Saudi Arabia's Sovereign Fund Will Also Invest $45B in SoftBank's Second Vision Fund," *TechCrunch*, October 2018, https://techcrunch.com/2018/10/07/saudi-arabias -sovereign-fund-will-also-invest-45b-in-softbanks-second-vision-fund/.

178 **Uber's multi-billion-dollar *quarterly* losses**: "Aramco Value to Top $2 Trillion, Less Than 5 Percent to Be Sold, Says Prince," *Reuters*, April 25, 2016, https://www.reuters.com/article/us-saudi-plan-aramco-idUSKCN0XM16M.

178 **the House of Saud**: Brooker, "The Most Powerful Person."

178 **thwart municipal officials' attempts at enforcement**: Mike Isaac, "How Uber Deceives the Authorities Worldwide," *New York Times*, March 3, 2017, https://www.nytimes.com/2017/03/03/technology/uber-greyball-program -evade-authorities.html.

179 **"Even if that means paying money"**: Dara Khosrowshahi, "The Campaign for Sustainable Mobility," Uber, September 26, 2018, https://www.uber.com/ newsroom/campaign-sustainable-mobility/.

180 **Five-cent nickel fares**: Cheape, *Moving the Masses*, 174–75.

180 **cities . . . grant a ride-hail monopoly**: "Free Exchange: The Market for Driverless Cars Will Head towards Monopoly," *The Economist*, June 7, 2018, https:// www.economist.com/finance-and-economics/2018/06/07/the-market-for -driverless-cars-will-head-towards-monopoly.

180 **"corrupt and contented"**: Cheape, *Moving the Masses*, 177.

180 **Jay Gould's Manhattan Railway Company**: Terry Golway, *Machine Made: Tammany Hall and the Creation of Modern American Politics* (New York: Liveright, 2014), 135.

180 **took over Puget Sound's streetcar**: Crowley, "City Light's Birth."

181 **Public transit was the competition**: United States Securities and Exchange Commission, *Registration Statement under the Securities Act of 1933: Uber*

Technologies, April 11, 2019, 25, https://www.sec.gov/Archives/edgar/data/1543151/000119312519103850/d647752ds1.htm#toc.

181 **deploy predatory pricing**: United States Securities and Exchange Commission, *Registration Statement*.

182 **"have been created based on cash flows"**: "Asset-Backed Security," Investopedia, accessed December 7, 2018, https://www.investopedia.com/terms/a/asset-backedsecurity.asp.

183 **Amazon's body-tracking technology**: "The Learning Machine: Amazon's Empire Rests on its Low-Key Approach to AI," *The Economist*, April 11, 2019, https://www.economist.com/business/2019/04/13/amazons-empire-rests-on-its-low-key-approach-to-ai.

8. Urban Machines

185 **"If enough people see the machine"**: John Heilemann, "Reinventing the Wheel," *Time*, December 2, 2001, http://content.time.com/time/business/article/0,8599,186660-3,00.html.

185 **Beijing is already on its seventh:** For an excellent look at the past, present, and future of China's new obsession with the automobile and its impact on urban design and planning, see Thomas Campanella, *The Concrete Dragon* (New York: Princeton Architectural Press, 2008).

186 **The push-button elevator**: Carlton Reid, "Driverless Vehicles Will Transform Cities? One Already Has: The Elevator," *Forbes*, February 18, 2019, https://www.forbes.com/sites/carltonreid/2019/02/18/driverless-vehicles-will-transform-cities-one-already-has-the-elevator/#255728e05be8.

186 **automatic traffic signals**: William Ghiglieri, Traffic signal, US Patent 1,224,632, filed January 12, 1915, and issued May 1, 1917, https://patentimages.storage.googleapis.com/99/ce/de/5ba1f0dd634ded/US1224632.pdf.

186 **to automate cities**: Geoffrey West, *Scale: The Universal Laws of Life, Growth, and Death in Organisms, Cities, and Companies* (New York: Penguin Random House, 2018).

186 **city dwellers worldwide expanded**: "World Population Increasingly Urban with More Than Half Living in Urban Areas," United Nations, July 10, 2014, http://www.un.org/en/development/desa/news/population/world-urbanization-prospects-2014.html.

189 **A "sky garage" car elevator**: Oliver O'Connell, "Pimp My Penthouse! Inside the $20 Million New York Apartment Boasting Its Own Car Elevator," *Daily Mail UK*, March 19, 2015, https://www.dailymail.co.uk/news/article-3001606/

Pimp-penthouse-Inside-20-million-New-York-apartment-boasting-CAR
-ELEVATOR.html.

190 **the clannish power structure**: Burak Arikan, "Networks of Disposses-
sion," Burak Arikan (website), 2013, https://burak-arikan.com/networks-of
-dispossession/.

190 **"clean the streets ... take care of plants"**: "Where Do Cars Go at Night?"
Moovel Lab, November 7, 2016, https://www.move-lab.com/projects/where
-do-cars-go-at-night.

190 **"As fracking upended the oil industry"**: Jack Sidders and Jess Shankleman,
"A Driverless Future Threatens the Laws of Real Estate," *Bloomberg*, February
6, 2018, https://www.bloomberg.com/news/articles/2018-02-06/a-driverless
-future-threatens-the-laws-of-real-estate.

191 **six billion parking stalls worldwide**: Patrick J. Kiger, "Designing for the
Driverless Age," Urban Land Institute, July 23, 2018, https://urbanland.uli.org/
planning-design/designing-driverless-age/.

191 **at nearly two billion**: Laura Bliss, "America Probably Has Enough Parking
Spaces for Multiple Black Fridays," *CityLab*, November 27, 2018, https://www
.citylab.com/transportation/2018/11/parking-lots-near-me-shopping-plazas
-vacant-spaces/576646/.

191 **public off-street parking spaces**: Matthew Flamm, "Driverless Cars
Could Let City Reclaim Parking Spots for Other Uses," *Crain's New York
Business*, July 12, 2017, http://www.crainsnewyork.com/article/20170712/
TECHNOLOGY/170719954.

191 **reclaimable parking lots**: Peter Madden, "Robots in Cities," *Huffington Post
UK*, September 11, 2017, https://www.huffingtonpost.co.uk/peter-madden
-obe/robots-in-cities_b_18504346.html?guccounter=1.

191 **parking garages will be obsolete**: Adele Peters, "These Future-Proof Parking Garages Can Easily Morph into Offices or Housing," *Fast Company*, January 14, 2019, https://www.fastcompany.com/90291136/these-futureproof
-parking-garages-can-be-easily-turned-into-offices-or-housing.

192 **parking structure's construction costs**: Alan Ohnsman, "The End of Parking Lots as We Know Them: Designing for a Driverless Future," *Forbes*, May
18, 2018, https://www.forbes.com/sites/alanohnsman/2018/05/18/end-of
-parking-lot-autonomous-cars/#15cfcdbe7244.

192 **Knightley's reopened in 2018**: Matt Riedl, "Living in a Parking Garage? These
Unique Wichita Lofts Are the First of Their Kind," *Wichita Eagle*, April 12,
2018, https://www.kansas.com/entertainment/ent-columns-blogs/keeper-of
-the-plans/article208445814.html.

193 **19 million parking spots consuming 200 square miles**: "Parking Is Real Estate in Hiding," *The Micromobility Newsletter*, April 23, 2019, https://micromobility.substack.com/p/parking-is-real-estate-in-hiding.

194 **"dynamics of future streets"**: National Association of City Transportation Officials [NACTO], *Blueprint for Autonomous Urbanism* (New York: NACTO, Fall 2017), 24.

194 **manage all kinds of movements**: NACTO, *Blueprint for Autonomous Urbanism*, 54–57.

194 **one's right to exist in safety**: NACTO, *Blueprint for Autonomous Urbanism*, 25.

197 **"rental costs account for just 5 percent"**: Hugh R. Morley, "Building Boom Not Likely to Ease NY-NJ Warehouse Rate Pressure," *JOC.com*, April 25, 2018. https://www.joc.com/port-news/us-ports/port-new-york-and-new-jersey/building-boom-not-likely-ease-ny-nj-warehouse-rate-pressure_20180425.html.

197 **eating out less than ever**: Sarah Kaufman, "Online Consumption and Mobility Practices: Crossing Views from Paris and Manhattan," NYU Wagner, November 26, 2018, https://wagner.nyu.edu/impact/research/publications/online-consumption-and-mobility-practices-crossing-views-paris-and.

197 **China's fastest-growing sectors today**: Mara Hvistendahl, "China's Tech Giants Want to Go Global. Just One Thing Might Stand in Their Way," *MIT Technology Review*, December 19, 2018, https://www.technologyreview.com/s/612598/chinas-tech-giants-want-to-go-global-just-one-thing-might-stand-in-their-way/.

199 **"sheds with beds"**: Citi GPS and Oxford Martin Programme on Technology and Employment, *Technology at Work v. 3.0: Automating E-commerce from Click to Pick to Door, Global Perspectives and Solutions* (Citigroup, 2017), 83.

202 **synergy of software trains and compact neighborhoods**: John Markoff, "Urban Planning Guru Says Driverless Cars Won't Fix Congestion," *New York Times*, October 27, 2018, https://www.nytimes.com/2018/10/27/technology/driverless-cars-congestion.html.

202 **Switzerland's Les Vergers Ecoquartier**: "Venir aux Verges," Les Vergers Ecoquartier, accessed May 23, 2019, https://www.lesvergers-meyrin.ch/ecoquartier/venir-aux-vergers.

202 *Micromobility*, **as this emerging sector is called**: Joe Cortright, "What Drives Ride-Hailing: Parking, Drinking, Flying, Peaking, Pricing," *City Commentary*, February 19, 2018, http://cityobservatory.org/what-drives-ride-hailing-parking-drinking-flying-peaking-pricing/; **under-five-mile journeys**: "Disrupting the Car: How Shared Cars, Bikes, Scooters, Are Reshap-

ing Transportation and Cannibalizing Car Ownership," *CB Insights Research Briefs*, September 5, 2018, https://www.cbinsights.com/research/disrupting -cars-car-sharing-scooters-ebikes/.

202 **more rides via its Jump bike-share**: Tony Bizjack, "What's More Popular Than Uber? Shockingly, Jump Bikes," *Sacramento Bee*, February 25, 2019, https://www.sacbee.com/news/local/article226640274.html.

202 **"disrupting" walking**: Portland Bureau of Transportation, *2018 E-scooter Findings Report* (Portland, OR: PBOT, 2018), 20.

205 **the vast periurban regions**: Terry G. McGee, "The Emergence of Desakota Regions in Asia: Expanding a Hypothesis," in *The Extended Metropolis: Settlement Transition in Asia*, ed. Norton Ginsburg et al. (Honolulu: University of Hawaii Press, 1991), 3–25.

206 **"bringing remote storage locations"**: "Automated Trucking: A CBRE Research Perspective," CBRE, November 17, 2017, http://www.cbre.us/real -estate-services/real-estate-industries/industrial-and-logistics/industrial -and-logistics-research/automated-trucking.

206 **"service firms should locate"**: CBRE, "Automated Trucking."

206 **"In order to feed, maintain, and entertain"**: Rem Koolhaas, "The World in 2018," *The Economist*, November 28, 2017, 153.

208 **"Rather than building AI"**: Russell Brandom, "Self-Driving Cars Are Headed Toward an AI Roadblock," *The Verge*, July 3, 2018, https://www.theverge .com/2018/7/3/17530232/self-driving-ai-winter-full-autonomy-waymo -tesla-uber.

209 **The term *jaywalking***: "Why Jaywalking Is Called Jaywalking," *Merriam-Webster*, accessed May 23, 2019, https://www.merriam-webster.com/words -at-play/why-is-it-called-jaywalking.

209 **weaponized by automobile interests**: Peter Norton, *Fighting Traffic: The Dawn of the Motor Age in the American City* (Cambridge, MA: MIT Press, 2011), 210.

209 **another separate guideway**: Kiger, "Designing for the Driverless Age."

209 **"argued for dense multilevel traffic"**: "New York Modern," The Skyscraper Museum, October 24, 2007, https://www.skyscraper.org/EXHIBITIONS/ FUTURE_CITY/new_york_modern.htm.

210 **"the first place in the world"**: Sidewalk Labs, *RFP Submission for Waterfront Toronto* (New York: Sidewalk Labs, 2017), 144.

210 **"Carriage squares"**: Renate van der Zee, "Story of Cities #30: How This Amsterdam Inventor Gave Bike-Sharing to the World," *The Guardian*, April 26, 2016,

https://www.theguardian.com/cities/2016/apr/26/story-cities-amsterdam
-bike-share-scheme.

210 **"a small zone that will serve"**: Sidewalk Labs, *RFP Submission for Waterfront Toronto*, 144.

211 **the "beeping monster"**: Shirley Zhao, "Tech Worries Throw Future of Hong Kong's First Driverless Electric Bus Route into Doubt," *South China Morning Post*, March 31, 2019, https://www.scmp.com/news/hong-kong/transport/article/3003944/driverless-electric-bus-fails-create-buzz-hong-kong.

211 **"Purchases are deposited at freight delivery centres"**: Sustainable Urban Mobility Research Laboratory, "Shared World," Singapore University of Technology and Design, accessed January 6, 2019, https://mobility.sutd.edu.sg/shared_world/.

211 **the unbuilt Minnesota Experimental City**: David Grossman, "The Time Minnesota Almost Built a Doomed, Future City," *Popular Mechanics*, March 31, 2018, https://www.popularmechanics.com/technology/infrastructure/a19642881/spilhaus-experimental-city-documentary/.

9. Wrestling with Regulation

213 **"The right to have access to every building"**: Lewis Mumford, *The Highway and the City* (San Diego, CA: Harcourt, Brace & World, 1963).

213 **One of the first to try**: "Toronto among the Fastest Growing Tech Hubs in North America," *U of T News*, July 21, 2017, https://www.utoronto.ca/news/toronto-among-fastest-growing-tech-hubs-north-america.

214 **potential safety, economic, and land use benefits**: David Ticoll, "Driving Changes: Automated Vehicles in Toronto," University of Toronto Transportation Research Institute Discussion Paper, October 15, 2015, https://munkschool.utoronto.ca/ipl/files/2016/03/Driving-Changes-Ticoll-2015.pdf.

214 **specific actions to align AV policy**: "Automated Vehicles Tactical Plan," Toronto, accessed September 10, 2019, https://www.toronto.ca/services-payments/streets-parking-transportation/automated-vehicles/draft-automated-vehicle-tactical-plan-2019-2021/.

214 **"Those who don't have automobiles"**: Lawrence D. Burns and Christopher Shulgan, *Autonomy: The Quest to Build the Driverless Car—and How It Will Reshape Our World* (New York: Ecco, 2018), 5.

215 **"I think public transport is painful"**: Aarian Marshall, "Elon Musk Reveals

His Awkward Dislike of Mass Transit," *Wired*, December 14, 2017, https://www .wired.com/story/elon-musk-awkward-dislike-mass-transit/.

215 **ride-hail's growth accompanied**: Michael Graehler Jr. et al., "Understanding the Recent Transit Ridership Decline in Major U.S. Cities: Service Cuts or Emerging Modes?" 98th Annual Meeting of the Transportation Research Board, November 14, 2018, https://usa.streetsblog.org/wp-content/uploads/ sites/5/2019/01/19-04931-Transit-Trends.pdf.

215 **bus ridership declined 20 percent**: Matt Tinoco, "Metro's Declining Ridership Explained," Curbed LA, August 29, 2017, https://la.curbed .com/2017/8/29/16219230/transit-metro-ridership-down-why.

215 **ride-hail's sapping effect on bus ridership**: Graehler et al., "Understanding the Recent Transit Ridership Decline."

216 **driverless city buses underway in Edinburgh**: "Edinburgh, UK," Initiative on Cities and Autonomous Vehicles, accessed September 10, 2019, https:// avsincities.bloomberg.org/global-atlas/europe/uk/edinburgh-uk.

216 **riders a high-tech experience**: "Nobina and Scania Pioneer Full Length Autonomous Buses in Sweden," Nobina, February 20, 2019, https://www .nobina.com/en/press/archive/nobina-and-scania-pioneer-full-length -autonomous-buses-in-sweden/.

216 **feature single-passenger self-driving pods**: "Toyota Partnership to Pilot Autonomous Vehicle Transportation System," *Nikkei Asian Review*, October 8, 2018, https://asia.nikkei.com/Business/Companies/Toyota-partnership-to -pilot-autonomous-vehicle-transportation-system.

216 **Combine Jelbi with a road-pricing platform**: David Zipper, "Bikeshare, Scooters, Cars, Trains, Bridges: One Agency to Rule Them All," *CityLab*, November 30, 2018, https://www.citylab.com/perspective/2018/11/transit -city-department-scootershare-ridehail-bikeshare/576982/.

217 **"the presence of an operator ensures"**: "Principles for the Transit Workforce in Automated Vehicle Legislation and Regulation," Transport Trades Department, March 11, 2019, https://ttd.org/policy/principles-for-the-transit -workforce-in-automated-vehicle-legislation-and-regulations/.

217 **"third space"**: Ray Oldenburg, *The Great Good Place: Coffee Shops, Bookstores, Bars, Hair Salons, and Other Hangouts at the Heart of a Community* (Cambridge, MA: Da Capo Press, 2000), 20.

217 **"a half-dozen startups are testing"**: Aarian Marshall, "Self-Driving Trucks Are Ready to Do Business in Texas," *Wired*, August 6, 2019, https://www.wired .com/story/self-driving-trucks-ready-business-texas/.

217 **encouraging night delivery**: Haag and Hu, "1.5 Million Packages a Day."

217 **how much freight moves**: Alain Bertaud, *Order without Design: How Markets Shape Cities* (Boston: MIT University Press, 2018), 30.

218 **road traffic in the developed world**: Bertaud, *Order without Design*, 30.

218 **number of store trips they take by as much as half**: Joann Muller, "One Big Thing: The Rise of Driverless Delivery," *Axios*, November 28, 2018, https://www .axios.com/autonomous-vehicles-could-be-used-for-deliveries-3fb12a24 -3e66-4d8b-b678-a2fbb47d05cb.html.

218 **more online shopping could empty them**: For an excellent overview of trends and interacting issues, see Joe Cortright, "Does Cyber Monday Mean Delivery Gridlock Tuesday?" *City Commentary*, November 29, 2016, http:// cityobservatory.org/cyber-monday-gridlock_tuesday/.

218 *one million* **parcel-hauling AVs**: KPMG, *Autonomy Delivers: An Oncoming Revolution in the Movement of Goods* (white paper, KPMG, 2018), 15.

219 **new lines of service**: "Autonomous Vehicles Could Be Huge for Small Businesses," *Axios*, November 16, 2018, https://www.axios.com/autonomous-vehi-cles-small-businesses-ford-ae0f4ea1-2aa9-44b7-bc64-a21f06b2eecb.html.

219 **hemorrhaged an estimated $1 billion**: Nanette Byrnes, "How Amazon Loses on Prime and Still Wins," *MIT Technology Review*, July 12, 2016, https:// www.technologyreview.com/s/601889/how-amazon-loses-on-prime-and -still-wins/.

219 **estimated to carry half its own shipments**: Syed M. Zubair Bokhari, "Amazon Now Its Own Largest Carrier," *Supply Chain Digest*, June 28, 2019, http:// www.scdigest.com/ontarget/19-06-28-1.php; **up from just 8 percent**: David Yaffe-Bellany and Michael Corkery, "FedEx Ends Amazon's U.S. Ground Deliveries as Retailer Rises as Rival," *New York Times*, August 7, 2019, https://www .nytimes.com/2019/08/07/business/fedex-amazon-shipping.html.

219 **consequences of a last-mile Amazon monopoly**: Stacy Mitchell, "Amazon Doesn't Just Want to Dominate the Market—It Wants to Become the Market," *The Nation*, February 15, 2018, https://www.thenation.com/article/ amazon-doesnt-just-want-to-dominate-the-market-it-wants-to-become -the-market/.

219 **more than 100 consumer brands**: Alex Moazed, "How Amazon's Marketplace Supercharged Its Private-Label Growth," *Inc.*, November 11, 2018, https:// www.inc.com/alex-moazed/what-brands-need-to-know-about-amazons -private-label-growth-how-to-respond.html.

221 **customers prefer a live delivery person**: Julian Allen, Maja Piecyk, and Marzena Piotrowska, *An Analysis of the Same-Day Delivery Market and Operations in the UK*, Technical Report CUED/C-SRF/TR012 (Westminster, UK: Univer-

sity of Westminster, November 2018); **company's package lockers**: McKinsey & Company, *Parcel Delivery: The Future of the Last Mile* (September 2016), 7.

221 **Porters could even help direct**: Joshuah K. Stolaroff, "Energy Use and Life Cycle Greenhouse Gas Emissions of Drones for Commercial Package Delivery," *Nature Communications* 9 (2018): 409, https://www.nature.com/articles/s41467-017-02411-5.pdf.

221 **"Money is only a tool"**: Ayn Rand, *Atlas Shrugged* (New York: Random House, 1957).

222 **automobile-related taxes, fees, and fines**: "Special Report: How Autonomous Vehicles Could Constrain City Budgets," The States and Localities, *Governing*, January 6, 2019, http://www.governing.com/gov-data/gov-how-autonomous-vehicles-could-effect-city-budgets.html.

222 **"The transition to AVs"**: Ticoll, "Driving Changes."

222 **27 percent of municipal revenue comes from parking fees**: M. W. Adler, S. Peer, and T. Sinozic, "Autonomous, Connected, Electric Shared Vehicles (ACES) and Public Finance: An Explorative Analysis," *Transportation Research Interdisciplinary Perspectives*, published ahead of print, September 28, 2019, http://dx.doi.org/10.1016/j.trip.2019.100038.

222 **"This likely represents the lower bound"**: International Transport Board of the OECD, *The Shared-Use City: Managing the Curb*, ITF Corporate Partnership Board Report, 2018, 58.

222 **"The American city is wasting"**: Henry Grabar, "Give the Curb Your Enthusiasm," *Slate Metropolis*, July 19, 2018, https://slate.com/business/2018/07/curb-space-is-way-too-valuable-for-cities-to-waste-on-parked-cars.html.

222 **"Cashing In on the Curb"**: Stephen Goldsmith, "Cashing In on the Curb," *Governing*, July 24, 2018, http://www.governing.com/blogs/bfc/col-smart-cities-data-curb-sidewalk-mobility-value.html.

223 **"It's the most valuable space that a city owns"**: Aarian Marshall, "To See the Future of Cities, Watch the Curb. Yes, the Curb," *Wired*, November 22, 2017, https://www.wired.com/story/city-planning-curbs/.

10. Pushing Code

225 **"The ghosts swarm"**: Rae Amantrout, "Unbidden," in *Versed* (Middletown, CT: Wesleyan University Press, 2009).

229 **"if it is legible"**: Kevin A. Lynch, *The Image of the City* (Cambridge, MA: MIT Press, 1960), 3.

229 **worked to reinforce the separation**: Lynch, *The Image of the City,* 66–68.

230 **Legible London, a massive wayfinding effort**: Jenny S. Reising, "Legible London," SEGD, November 26, 2009, https://segd.org/legible-london.

230 **people burn calories and spend at local shops**: Lilli Matson, "Wayfinding and Walking in London," *Transport for London*, May 2013, http://www.impacts .org/euroconference/vienna2013/presentations/London%20Walking%20 -%20Vienna%20May%202013.pdf.

230 **"[The city] is seen in all lights and weathers"**: Lynch, *The Image of the City*, 1.

231 **"the preprogrammed route was mapped"**: Tim Faulkner, "Driverless Little Roady Shuttle Hits a Few Speed Bumps," *ecoRI News*, August 3, 2019, https://www.ecori.org/transportation/2019/8/2/little-roady-shuttle-reaches -milestone-hits-speedbumps.

231 **couldn't even detect a crosswalk**: Ryan Stanton, "Can Self-Driving Pizza Delivery Cars Follow Ann Arbor's Crosswalk Law?" Ann Arbor, MLive, October 6, 2017, https://www.mlive.com/news/ann-arbor/index.ssf/2017/10/can_ self-driving_pizza_deliver.html.

231 **"classified [Elaine Herzberg] as something other"**: Ryan Felton, "Video Shows Driver in Autonomous Uber Was Looking Down Moments before Fatal Crash," *Jalopnik*, March 21, 2018, https://jalopnik.com/video-shows-driver-in -fatal-autonomous-uber-crash-was-l-1823970417.

231 **computer-vision algorithms**: Katyanna Quach, "Racist Self-Driving Scare Debunked, inside AI Black Boxes, Google Helps Folks Go with TensorFlow," *The Register*, March 10, 2019, https://www.theregister.co.uk/2019/03/10/ai_ roundup_080319/.

231 **marking of pavement to help orient drivers**: Bill Loomis, "1900–1930: The Years of Driving Dangerously," *Detroit News*, April 26, 2015, https://www .detroitnews.com/story/news/local/michigan-history/2015/04/26/auto -traffic-history-detroit/26312107/.

232 **dinged for excessive idling**: Andrew Small, "CityLab Daily: The Race to Code the Curb," *CityLab*, April 2, 2019, https://www.citylab.com/authors/ andrew-small/.

233 **"cities need to invest in the mapping in ways"**: Kevin Webb (@kvnwebb), "We're building smart ways to encode information like curb regulation," Twitter, November 30, 2018, 1:41 a.m., https://twitter.com/kvnweb/ status/1068485225607585798.

234 **"an exponential runaway beyond"**: Vernor Vinge, "The Coming Technological Singularity: How to Survive in the Post-Human Era," VISION-21 Symposium, NASA Lewis Research Center and Ohio Aerospace Institute, March 30–31, 1993, https://edoras.sdsu.edu/~vinge/misc/singularity.html.

234 **John von Neumann had raised**: John Brockman, *Possible Minds* (New York: Penguin Press, 2019), 8.

234 **"AI enthusiasts have been making claims"**: Vinge, "The Coming Technological Singularity."

234 **"I have set the date 2045"**: Christianna Reddy, "Kurzweil Claims That the Singularity Will Happen by 2045," *Futurism*, October 5, 2017, https://futurism .com/kurzweil-claims-that-the-singularity-will-happen-by-2045/.

235 **"machine learning" to disguise the true nature**: Chris Smith et al., "The History of Artificial Intelligence," University of Washington, December 2006, https://courses.cs.washington.edu/courses/csep590/06au/projects/his tory-ai.pdf.

235 **"a deep level of understanding"**: Rodney Brooks, "Post: [For&AI] The Origins of Artificial Intelligence," *Robots, AI, and Other Stuff* (blog), April 27, 2018, https://rodneybrooks.com/forai-the-origins-of-artificial-intelligence/.

235 **"Contemporary neural networks do well on challenges"**: Gary Marcus, "Deep Learning: A Critical Appraisal," New York University, accessed January 22, 2019, https://arxiv.org/pdf/1801.00631.pdf.

236 **"If . . . driverless cars should also disappoint"**: Marcus, "Deep Learning."

236 **Frey tallied his predictions of the jobs**: Carl Benedikt Frey and Michael A. Osborne, "The Future of Employment: How Susceptible Are Jobs to Computerisation?" University of Oxford, September 17, 2013, 60, https:// www.oxfordmartin.ox.ac.uk/downloads/academic/The_Future_of_Empl oyment.pdf.

237 **"decisive strategic advantage"**: Nick Bostrom, *Superintelligence: Paths, Dangers, Strategies* (Oxford, UK: Oxford University Press, 2014), 96.

237 **a serious and sober discussion about what**: Bostrom, *Superintelligence*, 96.

238 **more than $25 million since 2015**: Author's estimate based on figures published by Future of Humanity Institute, "Timeline of the Future of Humanity Institute," June 25, 2018, https://timelines.issarice.com/wiki/Timeline_of_ Future_of_Humanity_Institute.

238 **aspiring singleton's inevitable ambitions**: Bostrom, *Superintelligence*, 118. Experts will take note that what I describe here isn't that different from "mobility management" or "active travel management." Guilty as charged. The point here is to reframe these obtuse technical planning concepts for a broader audience.

241 **"stop routing vehicles"**: Ryan Randazzo, "A Slashed Tire, a Pointed Gun, Bullies on the Road: Why Do Waymo's Self-Driving Vans Get So Much Hate?" *Arizona Republic*, December 14, 2018, https://www.azcentral.com/story/money/

business/tech/2018/12/11/waymo-self-driving-vehicles-face-harassment -road-rage-phoenix-area/2198220002/.

241 **"illustrations that accompany the patent"**: Matt Day and Benjamin Romano, "Amazon Has Patented a System That Would Put Workers in a Cage, On Top of a Robot," *Seattle Times*, September 7, 2018, https://www .seattletimes.com/business/amazon/amazon-has-patented-a-system-that -would-put-workers-in-a-cage-on-top-of-a-robot/.

242 **"In the past, associates would mark out the grid of cells"**: Brian Heater, "Amazon Built an Electronic Vest to Improve Worker/Robot Interactions," *TechCrunch*, January 18, 2019, https://techcrunch.com/2019/01/18/amazon -built-an-electronic-vest-to-improve-worker-robot-interactions/.

242 **"pedestrians [should be] connected not detected"**: National Association for City Transport Officials [NACTO], *Blueprint for Autonomous Urbanism* (New York: NACTO, Fall 2017), 25.

243 **generate opportunities for meaningful livelihoods**: Charles Montgomery, *Happy City: Transforming Our Lives through Urban Design* (New York: Farrar, Straus, and Giroux, 2013).

245 **"The street finds its own uses for things"**: William Gibson, "Burning Chrome," *Omni*, July 1982.

246 **factor into the urban equation**: "FAA Drone Registry Tops One Million," *Transportation.gov*, January 10, 2018, https://www.transportation.gov/ briefing-room/faa-drone-registry-tops-one-million?xid=PS_smithsonian.

247 **"drone noise becomes thing that pushes cities"**: David King, email message to the author, December 8, 2018.

247 **100 million lines of computer code in the average car**: Doug Newcomb, "The Next Big OS War Is in Your Dashboard," *Wired*, December 3, 2012, https:// www.wired.com/2012/12/automotive-os-war/.

249 **"Two roads diverged in a wood, and I"**: Robert Frost, "The Road Not Taken," in *Mountain Interval* (1916).

Afterword

251 **"Nothing has spread socialistic feeling in this country more than the use of automobiles"**: Woodrow Wilson, "The Young Man's Burden" (speech at North Carolina Society of New York at the Waldorf-Astoria Hotel, February 27, 1906), https://www.fhwa.dot.gov/highwayhistory/wilson.pdf.

251 **"He saw now that you can't go home"**: Thomas Wolfe, *You Can't Go Home Again* (New York: Scribner, 2011), 704.

253 **Our geographic self-sorting**: Bill Bishop, *The Big Sort: Why the Clustering of Like-Minded America Is Tearing Us Apart* (New York: Mariner Books, 2009), 5.

253 **three of every four Americans drive**: Adie Tomer, "America's Commuting Choices: Five Major Takeaways from 2016 Census Data," Brookings Institute, October 3, 2017, https://www.brookings.edu/blog/the-avenue/2017/10/03/americans-commuting-choices-5-major-takeaways-from-2016-census-data/.

254 **California's infill capacity**: John Landis et al., "The Future of Infill Housing in California: Opportunities, Potential, and Feasibility," *Departmental Papers (City and Regional Planning)*, January 2006, https://repository.upenn.edu/cgi/viewcontent.cgi?referer=https://www.google.com/&httpsredir=1&article=1038&context=cplan_papers.

254 **eight million Americans now work from home**: Dan Kopf, "Slowly but Surely, Working at Home Is Becoming More Common," *Quartz*, September 17, 2018, https://qz.com/work/1392302/more-than-5-of-americans-now-work-from-home-new-statistics-show/.

Index

Page numbers in *italics* refer to illustrations, captions, and tables. Page numbers followed by *n* refer to footnotes.